METHODS FOR THE DETERMINATION OF VITAMINS IN FOOD

Recommended by COST 91

Edited by

G. BRUBACHER, W. MÜLLER-MULOT

F. Hoffmann-La Roche and Co. Ltd. Basle, Switzerland

and

D. A. T. SOUTHGATE

Food Research Institute, Norwich, UK

ELSEVIER APPLIED SCIENCE P.UBLISHERS
LONDON and NEW YORK

ELSEVIER APPLIED SCIENCE PUBLISHERS LTD
Crown House, Linton Road, Barking, Essex IG11 8JU, England

Sole Distributor in the USA and Canada
ELSEVIER SCIENCE PUBLISHING CO., INC.
52 Vanderbilt Avenue, New York, NY 10017, USA

British Library Cataloguing in Publication Data

Methods for the determination of vitamins in food: recommended by COST 91.
1. Vitamins
I. Brubacher, G. II. Müller-Mulot, W.
III. Southgate, D.A.T.
641.1'8   TX553.V5

ISBN 0-85334-339-X

WITH 3 TABLES AND 23 ILLUSTRATIONS

© ELSEVIER APPLIED SCIENCE PUBLISHERS LTD 1985

The selection and presentation of material and the opinions expressed in this publication are the sole responsibility of the authors concerned.

All rights reserved. No part of this publication may be reproduced, stored in a retrieval system, or transmitted in any form or by any means, electronic, mechanical, photocopying, recording, or otherwise, without the prior written permission of the copyright owner, Elsevier Applied Science Publishers Ltd, Crown House, Linton Road, Barking, Essex IG11 8JU, England

Photoset and printed by Interprint Ltd
Printed in Great Britain by Page Bros. (Norwich) Limited

d20 esk

TROPICAL DEVELOPMENT & RESEARCH
INSTITUTE — SIR

ACC	
CLASS	CU 577.16 :543
BULL	08 - 1985
CAT.	5 - 8 - 85

Preface

In the course of the project COST 91*, on the Effects of Thermal Processing and Distribution on the Quality and Nutritive Value of Food, it became clear that approved methods were needed for vitamin determination in food. An expert group on vitamins met in March 1981 to set the requirements which these methods must meet. On the basis of these requirements, methods were selected for vitamin A, β-carotene, vitamin B_1 (thiamine), vitamin C and vitamin E. Unfortunately, for vitamins B_2 (riboflavin), B_6 and D only tentative methods could be chosen, since the methods available only partially fulfilled the requirements set by the expert group. For niacin and folic acid some references only could be given because none of the existing methods satisfied these requirements, and for vitamin B_{12}, vitamin K, pantothenic acid and biotin it was not considered possible to give even references.

All methods were carefully described in detail so that every laboratory worker could use them without being an expert in vitamin assay. In October 1983 an enlarged expert group on vitamins approved the compilation of methods and approached a publishing house with a view to publication. The editors wish to thank Dr Peter Zeuthen, the leader of the project COST 91, for his interest in their work, and Mr G. Vos who served as a secretary of the project and who was responsible for the Commission of European Communities which translated the text into English, German and French. The editors are especially obliged to the 'Bundesamt für Bildung und Wissenschaft' of the Swiss Federation for its financial support and to the company F. Hoffmann-La Roche and Co. Ltd, Basle, for placing its infrastructure at their disposal. They also are

---

*COST 91 is a concerted action inaugurated by the organisation COST (Coopération Européenne dans le domaine de la recherche Scientifique et Technique) concerning research on effects of thermal processing and distribution on the quality and nutritive value of food.

very grateful to Professor Seher, Münster, Federal Republic of Germany, who agreed that the method for determination of individual tocopherols in oils and fats developed by the German Society for Fat Science (DGF) (DGF 'Einheitsmethode' F-II 4) could be integrated in the manual.

We are indebted also to Elsevier Applied Science Publishers Ltd, London, for their generous offer to publish the English version of the manual and to allow the Commission of European Communities to distribute the German or French version between the collaborators of COST 91 and 91 bis.

The manual is intended to be used by laboratory workers and scientists involved in research work on food technology, nutritional surveys, establishing food composition tables and the quality control of food.

There are many gaps which should be closed in a second edition of the manual. For example, the missing data for precision and other characteristics should be elaborated, the methods should be compared with established methods and the limits of application should be investigated in collaborative trials. Unfortunately this work cannot be continued under the patronage of COST 91, but all those concerned with vitamin analysis are urged to arrange corresponding trials and to refer the results to the editors. It is hoped that the editors will find a new organisation to take over the patronage.

The editors hope that the present manual will be of help to many of their colleagues and wish to thank all those colleagues who have contributed to its completion.

G. BRUBACHER
W. MÜLLER-MULOT
D.A.T. SOUTHGATE

Contents

Preface v

Members of the Expert Group on Vitamin Determination ix

PART I: INTRODUCTION

1. Introduction 3

PART II: RECOMMENDED METHODS

2. Vitamin A (Retinol and Retinyl Esters) in Food: HPLC Method 23
3. Carotene in Foodstuffs.. 33
4. Vitamin B_1 (Thiamine) in Foodstuffs: Thiochrome Method... 51
5. Vitamin C (Ascorbic and Dehydroascorbic Acids) in Foodstuffs: HPLC Method 66
6. Vitamin C (Ascorbic and Dehydroascorbic Acids) in Foodstuffs: Modified Deutsch and Weeks Fluorimetric Method 76
7. Vitamin C (Ascorbic and Dehydroascorbic Acids) in Food: Sephadex Method 85
8. Vitamin E (Only α-Tocopherol) in Foodstuffs: HPLC Method   97
9. Free Tocopherols and Tocotrienols (Vitamin E) in Edible Oils and Fats: HPLC Method....................................... 107

PART III: TENTATIVE METHODS

10. Vitamin B_2 (Riboflavin) in Foodstuffs: HPLC Method 119

11. Vitamin B_6 in Foodstuffs: HPLC Method..................... 129
12. Vitamin D in Margarine: HPLC Method 141

PART IV: ANNEX

13. Niacin... 155
14. Folacin in Foodstuffs 158

Index... 161

Members of the Expert Group on Vitamin Determination

Prof. Dr A. E. BENDER
2 Willow Vale
Fetcham, Leatherhead
Surrey KT22 9 TE
UK

Dr A. BLUMENTHAL
Institut für Ernährungsforschung
Seestrasse 72
CH-8803 Rüschlikon
Switzerland

Dr rer. nat. A. BOGNÁR
Bundesforschungsanstalt für
 Ernährung
Institut für Ernährungsökonomie
 und -soziologie
Garbenstrasse 13
D-7000 Stuttgart 70
Federal Republic of Germany

Dr F. BRAWAND
Schweizerisches Vitamininstitut
Vesalianum
Vesalgasse 1
CH-4051 Basel
Switzerland

Prof. Dr G. B. BRUBACHER
Department of Vitamin and
 Nutrition Research
F. Hoffmann-La Roche & Co. Ltd
Grenzacherstrasse 124
CH-4002 Basel
Switzerland

Prof. Dr F. FIDANZA
Istituto di Scienza dell' Alimenta-
 zione
Università degli Studi di Perugia
Casella Postale 333
I-06100 Perugia
Italy

Dr U. MANZ
F. Hoffmann-La Roche & Co. Ltd
Grenzacherstrasse 124
CH-4002 Basel
Switzerland

Dr W. MÜLLER-MULOT
Bürgelerweg 15
D-7889 Grenzach-Wyhlen 1
Federal Republic of Germany

MEMBERS OF THE EXPERT GROUP

Dr P. SHEFFELDT
Institut für Ernährungsforschung
Seestrasse 72
CH-8803 Rüschlikon
Switzerland

Dr W. SCHÜEP
F. Hoffmann-La Roche & Co. Ltd
Grenzacherstrasse 124
CH-4002 Basel
Switzerland

Mr D. SCUFFAM
Laboratory of the Government Chemist
Department of Industry
Cornwall House
Stamford Street
London SE1 9NQ
UK

Dr D. A. T. SOUTHGATE
Food Research Institute
Agricultural and Food Research Council
Colney Lane
Norwich NR4 7UA,
UK

Dr S. SREBRNIK
Institut d'Hygiène et d'Epidémiologie
rue Juliette Wytsman 14
B-1050 Bruxelles
Belgium

Dr J.-P. VUILLEUMIER
F. Hoffmann-La Roche & Co. Ltd
Grenzacherstrasse 124
CH-4002 Basel
Switzerland

Prof. Dr P. WALTER
Schweizerisches Vitamininstitut
Vesalianum
Vesalgassse 1
CH-4051 Basel
Switzerland

Dr C. E. WEST
Co-ordinator of the Eurofoods Project
Department of Human Nutrition
Agricultural University
De Dreijen 12
6703 BC Wageningen
The Netherlands

PART I

INTRODUCTION

1
Introduction

The present manual on methods for determination of vitamins in food has been written mainly for practical purposes. It consists of a compilation of methods, which have been used successfully in the hands of experienced experts, and which have been chosen according to certain criteria given below. Unfortunately information on all criteria was not available for any of the methods. Therefore the methods were chosen by a consensus of the expert group where this information was missing. The lack of information is noted in the descriptions of the methods. It is hoped that in a second edition these gaps will be closed.

The description of the method consists of the following sections.

1. Purpose and Scope
2. Definition

These two sections should be noted carefully, since many misunderstandings have occurred in the past when these fundamental considerations have been disregarded. Some criteria on these two points used in choosing the method will be discussed later.

3. Brief description of the method (principle of the method)

The description gives a short overview of the method and allows the reader to decide whether a method may be performed with the available equipment in a certain laboratory or not.

4. Chemicals
5. Apparatus and Accessories

In sections 4 and 5 all chemicals and apparatus which are used for performing the actual determination are enumerated. This means that the description of the procedure refers only to these reagents and equipment. It does not mean that other similar chemicals and equipment cannot be used, but that the laboratory workers may need to modify the method accordingly.

It was decided not to use the complexity of equipment nor the availability of chemicals as selection criteria. But it is open to discussion as to whether or not methods which can be performed with less sophisticated equipment and chemicals should be included in a second edition.

    6. Sample, sampling and preparation of the sample for the laboratory

This is one of the most critical stages in food analyses. Originally it was the intention to choose only methods for which enough information was available on these matters. Unfortunately there was little or no information available. It was therefore decided not to include this as a criterion for selection, but to include a discussion on these points in this chapter.

    7. Procedure

This section is the major part of the description of each method. Every step is described in detail so that every competent laboratory worker should be able to perform the determination without special knowledge of vitamin assay techniques.

    8. Evaluation

This part contains the formula for calculating the result with the aid of the measured values and the criteria by which the results should be judged. These are also the main criteria by which the selection of the methods were made—namely precision, accuracy, sensitivity, specificity and robustness. A discussion of these terms is given below.

    9. Analysis report
    10. References.

Where some critical points are discussed, notes are inserted in the text, a comment on how to prepare the report is made and some actual references related to the method in question are given.

In addition to the description of well-established methods (vitamin A, β-carotene, vitamin B_1 (thiamine), vitamins C and E), tentative methods are also similarly described (vitamins B_2, B_6 and D) which have been tested with only a few food items or which are not sufficiently sensitive to cover the whole concentration range. For folic acid and niacin only references are given, since it was felt that time is not ripe to include detailed description of methods for these two vitamins. Finally there are neither methods nor references for determination of vitamin B_{12}, biotin, pantothenic acid and vitamin K. It is hoped that in the next edition vitamin B_{12} will also be included.

PURPOSE, SCOPE AND DEFINITION

Vitamin determinations in foods are carried out for many purposes, e.g. in food technology it is desirable to know the fate of vitamins during processing, and in nutritional surveys the vitamin content of meals at the point of consumption frequently has to be measured. The vitamin content of food has also to be determined for establishing food composition tables or for legal purposes in connection with nutritional labelling. All these purposes have different criteria for selecting the most appropriate method. The main purpose of the present book is to give food technologists an instrument for investigating the fate of vitamins during processing.

In the vitamin assay of foodstuffs it must be kept in mind that the term 'vitamin' is a physiological one rather than a chemical one, expressing a certain physiological activity, which is related to the chemical substances which are responsible for this activity. In this connection the following two points have to be considered.

(a) Vitamers

Vitamin activity may be due to a group of different chemical compounds (vitamers). These chemical compounds can be divided into two classes: (i) compounds which can be easily converted by simple chemical or biochemical reactions into the active form, e.g. vitamin A palmitate may be saponified to retinol, or dehydroascorbic acid may be reduced to ascorbic acid; (ii) compounds which cannot be inter-converted by simple means, e.g. α-, β-, γ- and δ-tocopherol, or α- and β-carotene. In Figs. 1–9 the most common compounds which are found naturally in foods or which are added during processing, are given for vitamins A, B_1, B_2, niacin, and vitamins B_6, C, D, E and folic acid.

A comprehensive method would allow determination of each chemical compound separately. In this way the fate of each compound during storage and food processing could be followed. Unfortunately, no practicable comprehensive methods exist. Only in special cases, usually in connection with research on biological problems, has such an approach been realised by developing special analytical procedures. For practical

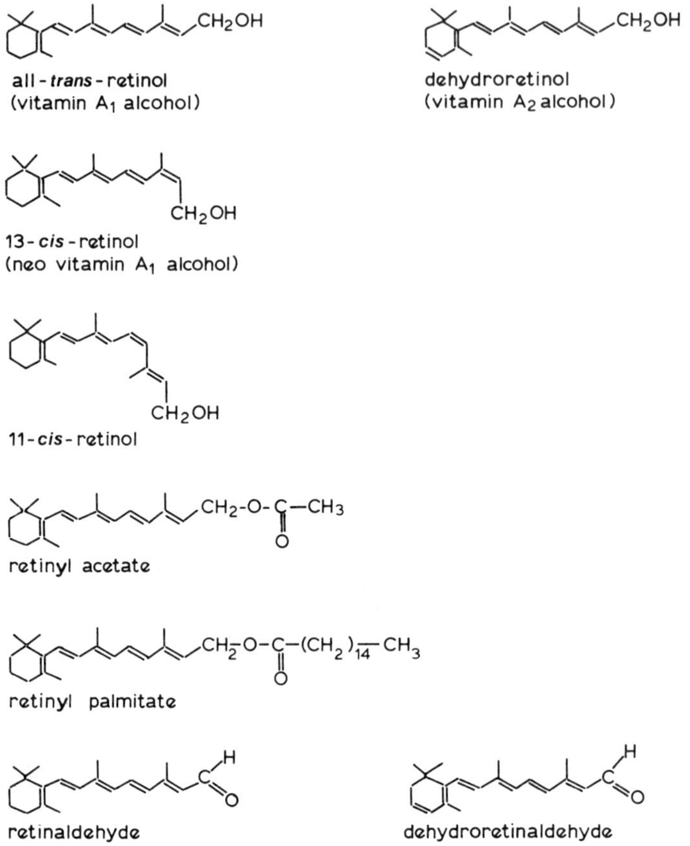

Fig. 1. Most common compounds with vitamin A activity (vitamers of the vitamin A group).

purposes a more pragmatic approach has to be adopted. Methods are selected, where the main interconvertible vitamers are converted to the same vitamer in the course of the analysis. For example, in the method recommended for vitamin A, retinyl palmitate and retinyl acetate are converted to retinol and the result of the analysis is given as weight units of retinol per 100 g. It is not possible to derive from this result the original quantity of retinol, retinyl palmitate and retinyl acetate. If one is following the fate of vitamin A during storage and processing, and at the end of the experiment a lower value is found for retinol than at the

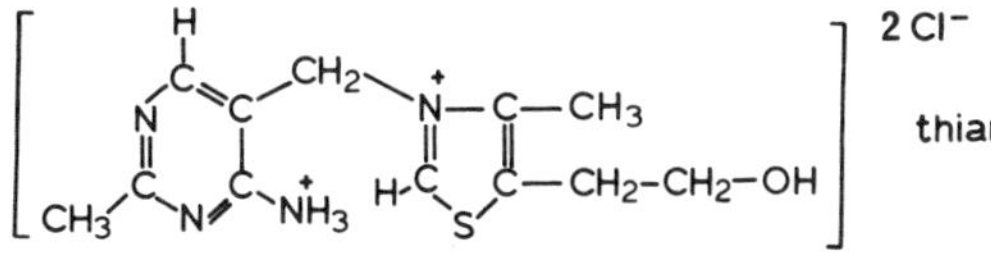

thiamine chloride hydrochloride

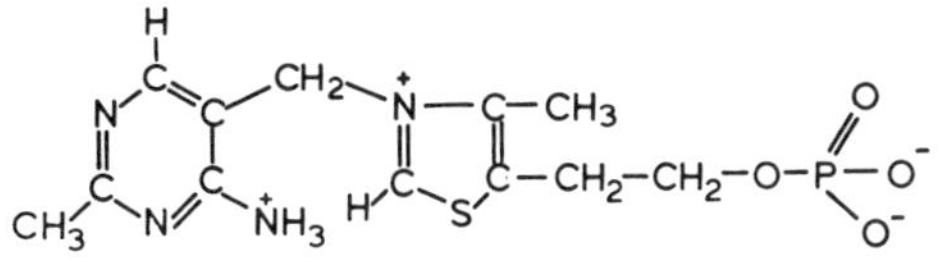

thiamine monophosphate

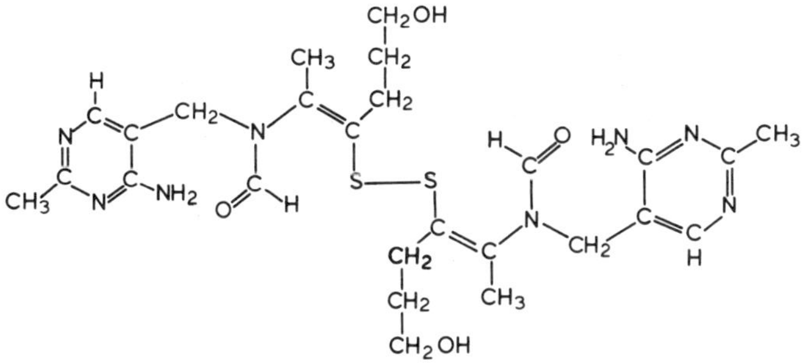

thiamine disulphide

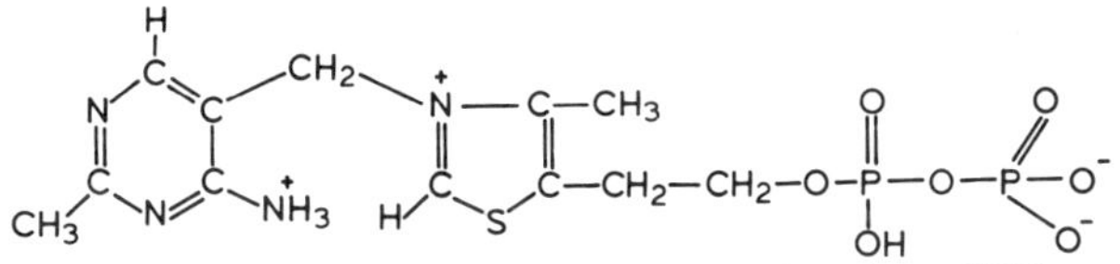

thiamine diphosphate, thiamine pyrophosphate (TPP)

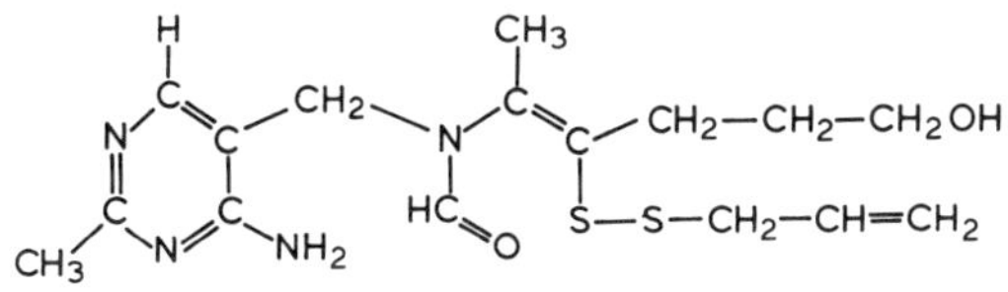

allyl thiamine disulphide

Fig. 2. Most common compounds with vitamin B_1 activity (vitamers of the vitamin B_1 group).

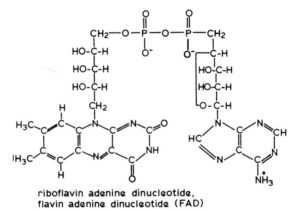

Fig. 3. Most common compounds with vitamin B_2 activity (vitamers of the vitamin B_2 group).

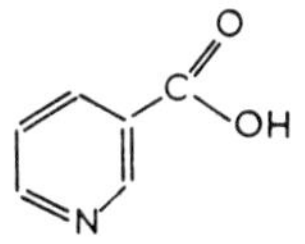

nicotinic acid

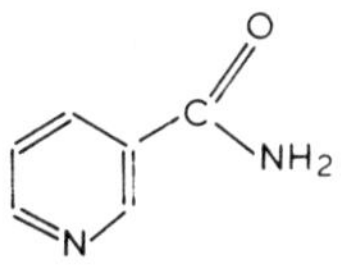

nicotinic acid amide
(nicotinamide, niacinamide)

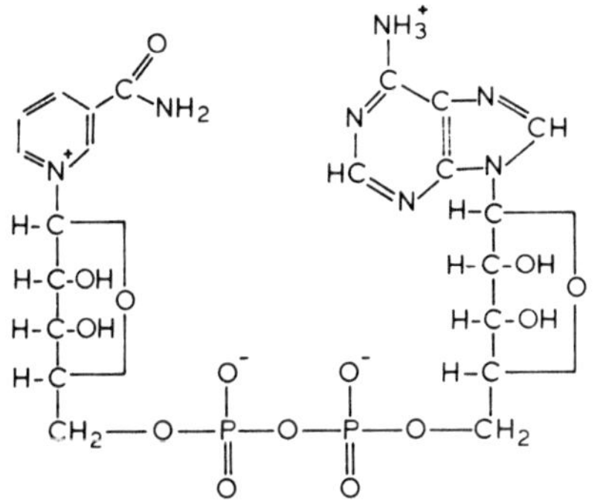

nicotinamide adenine dinucleotide (NAD)

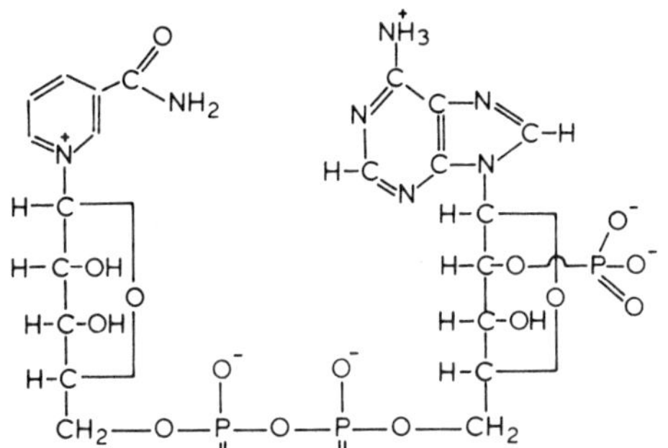

nicotinamide adenine dinucleotide phosphate (NADP)

Fig. 4. Most common compounds with niacin activity (vitamers of the niacin group).

beginning it can be concluded that some vitamin A was destroyed during the experiment. But if the same value is found at the start and after processing it is not possible to decide whether all vitamin A active compounds were stable during the experiment or whether some vitamin A palmitate or acetate has been transformed to retinol. It has also to be

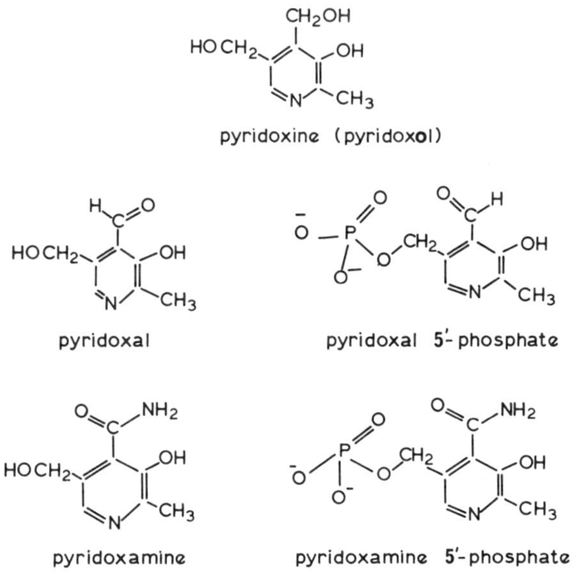

pyridoxine (pyridoxol)

pyridoxal

pyridoxal 5'-phosphate

pyridoxamine

pyridoxamine 5'-phosphate

Fig. 5. Most common compounds with vitamin B_6 activity (vitamers of the vitamin B_6 group).

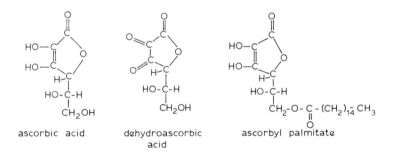

ascorbic acid

dehydroascorbic acid

ascorbyl palmitate

Fig. 6 Most common compounds with vitamin C activity (vitamers of the vitamin C group).

noted that the result refers only to vitamin A active compounds which are converted by the analytical procedure to retinol. The recommended method discounts, for instance, retinaldehyde but as the concentration of retinaldehyde in common foods is usually small compared to the concentration of retinol and retinyl esters, this procedure introduces only a

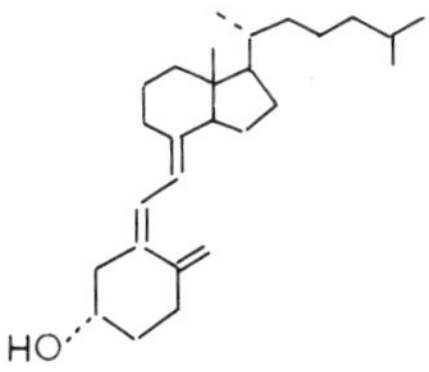

cholecalciferol
(vitamin D₃)

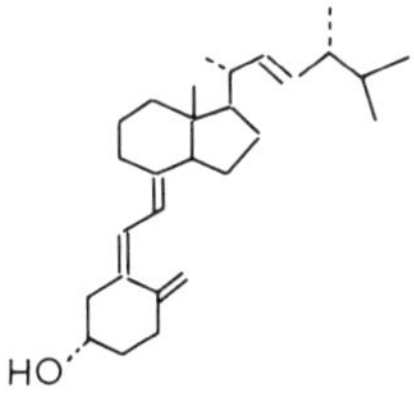

ergocalciferol
(vitamin D₂)

Fig. 7. Most common compounds with vitamin D activity (vitamers of the vitamin D group).

tocopherols:

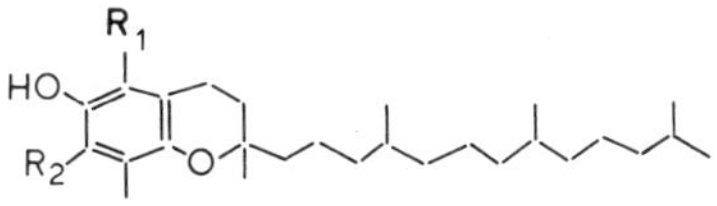

R_1	R_2		
CH_3	CH_3	α - tocopherol	(α - T)
CH_3	H	β - tocopherol	(β - T)
H	CH_3	γ - tocopherol	(γ - T)
H	H	δ - tocopherol	(δ - T)

tocotrienols:

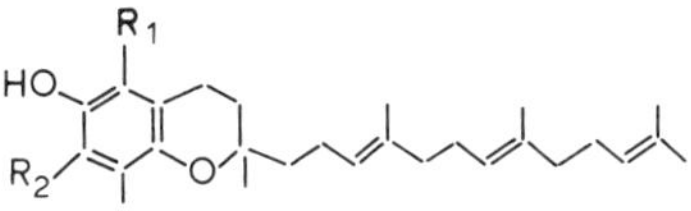

R_1	R_2		
CH_3	CH_3	α - tocotrienol	(α - T_3)
CH_3	H	β - tocotrienol	(β - T_3)
H	CH_3	γ - tocotrienol	(γ - T_3)
H	H	δ - tocotrienol	(δ - T_3)

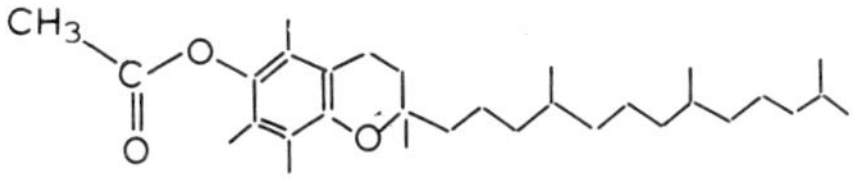

α - tocopherol acetate

Fig. 8. Most common compounds with vitamin E activity (vitamers of the vitamin E group).

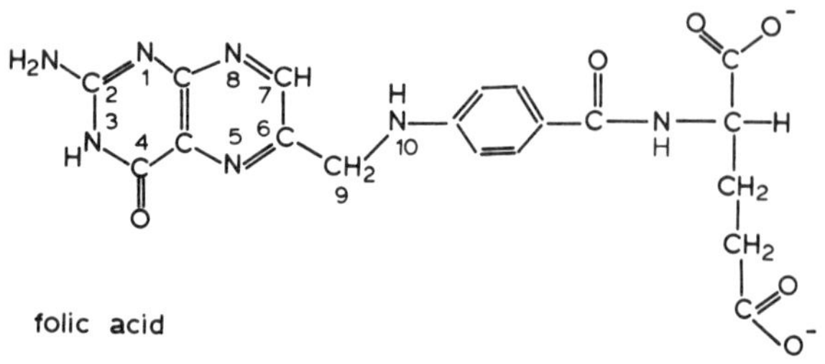

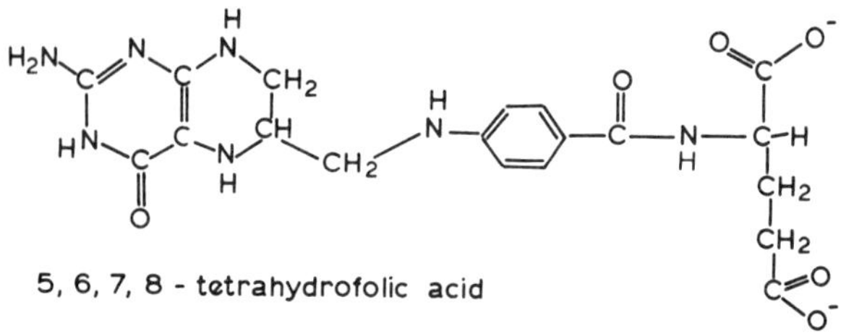

Fig. 9. Folate vitamins (vitamers of the folacin group). Food folates are usually derivatives of 5,6,7,8-tetrahydrofolic acid with glutamyl side chains of up to eight glutamyl residues. The majority have single carbon substituents, usually 5-methyl or 5-formyl.

small error. This may not be true in rare cases of special substrates where the application of the method would lead to false conclusions.

In cases where it is not possible to interconvert the vitamers to the same compound two alternative solutions are possible: either all vitamers are determined together (disregarding the individual vitamers) or the principal vitamer(s) has to be determined individually. For instance the method tentatively recommended for vitamin D does not distinguish between vitamins D_2 and D_3 and thus gives the sum of these two vitamers, whereas in the method recommended for the determination of vitamin E in foods only α-tocopherol is determined. This vitamer is the major form in most foods and has the highest biological activity.

In cases where food analysis is performed in connection with nutritional surveys or for establishing food composition tables it is absolutely necessary to indicate the method and to specify the compounds which have been measured. If, for instance, only one vitamer is determined (as is often the case for vitamin E in food) the result of the analysis gives a value which is too low for the total vitamin E activity of the food. If, on the other hand, all vitamers are determined together or converted to the same compound the nutritional value may be overestimated. This is the case, for instance, for bound niacin in cereals. Bound niacin may be converted to free niacin by saponification with alkali; however, bound niacin has only a low biological activity and would best be discounted.

In circumstances where analysis has to be carried out for legal purposes the legal regulations frequently define which substances are included in the definition of the vitamin activity and the methods that should be used, otherwise the producer or distributor would live in legal insecurity. In countries where official methods exist the methods in the present book can only be used if they can be shown to give the same results as the official methods, and in some cases only the official method is accepted. In countries where no such methods exist the methods may be suitable as proposed methods for the authorities to be considered as official methods.

(b) Biological Activity and Bioavailability

Even in cases where a compound may be converted into another, the biological activity on a molecular basis may not be identical. We have mentioned above the biological activity of bound niacin compared to free niacin: other examples are the different biological activities of all-*trans*-retinol and 13-*cis*-retinol or of precholecalciferol and cholecalciferol. In the ideal case all vitamers of a certain vitamin should be assayed separately and the result for each vitamer should be multiplied by its biological activity. Unfortunately this procedure is not feasible since no methods exist which permit the determination of each vitamer separately and the biological activity in man of many vitamers is not known. Furthermore, bioavailability depends not only on the chemical nature of the vitamers considered but also on the composition of the meals. For instance, β-carotene in raw carrots has only a low vitamin A activity, whereas the same compound in oily solution is almost completely transformed to vitamin A. It is clear that under such circumstances biological activity has to be disregarded in the description of the assay method. For most analytical purposes, corresponding to the main object of the present book, or where the methods are used for legal purposes, this missing information is of little concern, but in all other cases, where the results are used to estimate the nutritional value of food, the results have to be interpreted by an expert in nutrition, otherwise very erroneous conclusions could be drawn.

SAMPLE, SAMPLING AND PREPARATION OF THE SAMPLE FOR THE LABORATORY

In cases where the results of the vitamin determination are used for establishing food composition tables the sampling procedure is the most critical point. In technological experiments the sample is defined by the design of the experiment; in cases where the analysis is done for legal purposes, the method of sampling is usually prescribed by law. In all cases the pretreatment for conservation and transport to the laboratory is extremely important since most vitamins are very labile and could be easily lost in the interval between sampling and performing the analysis. However, little guidance is given in the literature on this topic. With regard to sampling of food items, some guidelines are given by Southgate in Ref. 1.

Ideally the food sample should be analysed immediately after selection. This can be done when the experiments are performed in the laboratory itself. Usually, however, the sample has to be handled outside the laboratory. In this case it has to be pretreated in a manner such that no destruction occurs during transportation to the laboratory and during storage. No general procedure can be given because, unfortunately, this matter has not been investigated thoroughly. In Ref. 2, Osborne and Voogt recommend the following procedure:

1. All bones and other non-edible materials are removed (noting the weight proportion of non-edible to edible parts).
2. The sample is frozen as quickly as possible at a temperature of $-20\,°C$ and stored at $-20\,°C$ until homogenisation. There is little or no information on how long the frozen sample can be stored without damaging the most labile vitamins.
3. The frozen sample is broken into small pieces with a wooden mallet or a pestle.
4. A representative sub-sample of about 500 g is taken.
5. The sub-sample is passed through a meat mincer (cooled to about $-20\,°C$) and transferred into a Dewar vessel filled with liquid nitrogen.
6. When the sub-sample has reached the temperature of the liquid nitrogen it is transferred to the beaker of a Waring blender cooled with liquid nitrogen.
7. The sub-sample is homogenised until a homogeneous powder is obtained.

8. The powder is transferred to precooled ($-20\,°C$) widemouthed sample bottles.
9. The sample bottles are stored in a plastic container in a freezer at $-20\,°C$ until required for analyses. (Samples for vitamin C analyses cannot be stored in this way and have to be analysed immediately after homogenisation.)

For vitamin C assay the samples of food, meals or beverages are best handled following the procedure of Vuilleumier (Ref. 3). They are frozen immediately after selection using dry ice and transported under dry ice as soon as possible to the laboratory.

In the laboratory the samples should never be thawed completely and hot water should never be used for thawing. Homogeneous samples (bread, biscuits, fruits, meat, milk products, potatoes, vegetables, etc.) are roughly diced by a stainless steel knife and weighed in a 250 ml polyethylene bottle. The bottle is gassed with carbon dioxide and metaphosphoric acid is added. The amount and concentration of metaphosphoric acid depends on the water content of the sample (see Table 1).

TABLE 1
Weight and Concentration of Metaphosphoric Acid for Vitamin C Assay

Item	Weight of the sample (g)	Weight of metaphosphoric acid (g)	Concentration of metaphosphoric acid (%)
Soups Beverages Fruits with sugar Salads	50	50	10
Meats Sausages Fruits Salads Mashed potato	30	70	10
Cooked potato Vegetables Bread Curd	20	80	5
Potato—chips Potatoes (french fried)	10	90	5

Samples which are inhomogeneous have to be thawed in a water bath at room temperature until they are semi-liquid, mixed by shaking and immediately weighed into the bottle.

The contents of the bottle are homogenised for 30 s using a high-frequency top-driven disintegrator. The mixture is allowed to stand for several minutes and centrifuged; if an oily phase is formed this phase has to be removed and the remainder is filtered through a filter paper into a 100 ml polyethylene bottle. The tightly-sealed bottle can be stored for 1–3 months at $-25\,°C$ to $-30\,°C$ with minimal loss of vitamin C.

To prepare 10% metaphosphoric acid solution, 200 g (or for 5% solution, 100 g) are dissolved in 1800 (or 1900) ml cold water without heating. The solution is filtered if necessary and can be stored in the refrigerator for two weeks.

In all cases where it is not possible to analyse the samples immediately after selection the sample should be frozen immediately after selection in dry ice (solid CO_2) and transported frozen to the laboratory and stored at a low temperature (between $-20\,°C$ and $-30\,°C$). The sample should be analysed as soon as possible. It is important not to reduce the sample to smaller pieces in order to keep the exposed surface area low. No systematic investigation has been done on the relationship between storage time and vitamin loss under this storage condition.

EVALUATION

The main criteria by which a measured value is judged are precision and accuracy (or correctness). For interpretation of these two concepts a model is used in which it is supposed that there exists a correct value which represents the true vitamin content of the sample (for details of this model see Ref. 4). If it was possible to perform an ideal analysis the result of such an analysis would differ from the true value by the so-called systematic error (sometimes called bias or constant error). Since no person can ever perform an ideal analysis the result of an actual analysis differs from the ideal result by the so-called statistical or random error. An estimate of the ideal value consists of the average of the actual results of a series of actual analyses performed on the same sample by the same person with the same equipment in the same laboratory. An estimation of the statistical error is the standard deviation of this set of results.

If the set of assays is performed by the same person with the same equipment in the same laboratory, one speaks of its 'repeatability'. Where different persons, different equipment or different laboratories are involved, one speaks of its 'comparability', and in both cases its 'reproducibility'. Thus accuracy (correctness) is high if the systematic error is small and the precision is high if the statistical error is small. It is important to realise that there is no method of determining the true value of the concentration of a component in a natural product, therefore there is no method of measuring the systematic error. In most of the procedures which are used, for example maximising the analytical signal, the use of an internal standard or of recovery tests can give at most a hint of whether an appreciable systematic error exists.

The only method which can be used for estimation of the accuracy of a value is by comparison of various independent methods. It is important, however, that the methods are truly independent. If two independent methods give the same result it is more probable that the measured value is near to the true value than that both methods have the same systematic error by chance. The probability of the measured value being close to the true value is even higher if three independent methods are used and all give the same result. On the other hand, if the results of two independent methods differ considerably at least one or both may have a considerable systematic error. Of the constituents covered in the present book only in the case of vitamin C have three methods been compared over a series of various food items (see Ref. 5) and have given more or less the same results; thus it can be concluded that the results are correct and that the accuracy of the three methods proposed is high. Only in case of high specificity can it be assumed that the accuracy of the method is constant, otherwise the systematic error is influenced by other components in the food and differs for various food items. In respect of precision it should be recognised that analytical results are in general not distributed normally and that therefore the classical statistical analysis of the problem has its limits. Nevertheless it is the only practicable way to proceed.

The precision of a method depends on the nature of the sample and on the concentration of the substance to be analysed. It is therefore important to indicate the field of application and the concentration range which corresponds to a given precision. Under optimal conditions there is usually a concentration range where the absolute value of the standard deviation increases linearly with the concentration; the relative standard deviation on the other hand remains more or less constant. With less optimal conditions and a lower concentration range the absolute standard deviation remains constant, whereas the relative standard deviation increases inversely to the concentration and, at even lower concentration the absolute and relative standard deviations increase inversely to the concentration. The detection limit of the method corresponds to the concentration where the relative standard deviation exceeds a certain value, in general about 25–50%.

For practical purposes it is very important that a proposed method has a high robustness or ruggedness, which means that small changes in the conditions of the analytical procedure (pH, temperature, purity of the analytical reagents, storage time of the sample, wavelength of measurement, etc.) result in insignificant changes in the measured results. A measure of the robustness of a method is its reliability, as this implies that the variation of the systematic and random errors with time and the number of extreme values which have to be excluded for statistical treatment of the results are low, and that there is good comparability between the results of different laboratories. No systematic investigations of robustness have been made for the recommended methods.

Depending on the purpose of the investigation the various analytical procedures have to satisfy different requirements. In all cases, where routine work is involved a high robustness is important. In cases where the fate of a certain vitamin during technological processing is investigated a high precision and high reliability are demanded, whereas accuracy is not so important because all results can be expressed as relative values.

In cases where the analysis is performed in connection with a nutritional survey the method should be very accurate and precise, since the absolute value is needed for further calculation. On the other hand the limit of detection should permit the assay of foodstuffs containing about 5–10% of an adult person's vitamin requirement per daily portion—smaller amounts can usually be neglected. The limit of detection is therefore not the limiting factor of the method in most cases.

In cases where the analysis serves for establishing food composition tables, high accuracy is required but the precision may be of secondary importance because the variance of the final figures derives mainly from the variance of the vitamin contents of the food items and not from the precision of the analytical procedure.

In cases of official analysis the accuracy is of secondary importance because it is rarely recognised that by definition the result of an ideal analysis is identical with the true value. The method to be used has therefore to be defined by law. A high precision of the method is advantageous and the limit of detection has to permit the assay of food with the lowest concentration which is foreseen in the law.

As was stated in the introduction for most of the methods recommended by the expert group, the data on accuracy, precision, etc., are missing. It is therefore incontestable that these characteristics have to be determined if the manual shall serve in a wider context. It is, however, not possible to give detailed advice on how these missing values should be obtained since each method poses its own problems; however, some general suggestions can be made.

First, it is desirable that as many analysts as possible should become acquainted with the proposed method. They should compare the results of the proposed method with the results of the method used in their own laboratories and, if possible also with the results of an official method and of other independent methods. They should also determine the precision of the method. It is better to estimate the precision by performing a series of duplicate determinations on a range of food items within a broad concentration range than a series of determinations on the same food item. Robustness should also be checked by systematic variation of as many variables as possible (pH, temperature, etc.). The observations made at this point should be referred to the editors of the manual and any necessary modifications should be proposed. Secondly, two or more analysts should compare the results which they obtain by assaying the same sample, and refer their observations to the editors again with proposals for modifications. At this stage collaborative trials may be necessary. Because collaborative assays are cumbersome and time consuming a second edition of the manual in which the proposed modifications are taken in consideration, may be more useful and collaborative trials should perhaps be only considered if the methods of the manual are adopted by official bodies. For planning the execution of the necessary trials Refs 6–11 are recommended.

REFERENCES

1. D. A. T. Southgate, *Guide Lines for the Preparation of Tables of Food Composition*, S. Karger, Basel, 1974.
2. D. R. Osborne and P. Voogt, *The Analysis of Nutrient in Foods*, Academic Press, London, 1978.
3. J.-P. Vuilleumier, Analytische Probleme bei der Bestimmung von Vitamin C im Zusammenhang mit Ernährungserhebungen. *Int. Z. Vit. Forschung*, **37**, 504–14 (1967).

4. William G. Schlecht, The probable error of a chemical analysis. *Contribution to Geochemistry Bulletin*, **992**, 57–69 (1949).
5. A. Floridi, R. Coli, A. Alberti Fidanza, C. F. Bourgeois and R. A. Wiggins, High-Performance Liquid Chromatographic Determination of Ascorbic Acid in Food: Comparison with other Methods. *Int. J. Vit. Nutr. Res.*, **52**, 194–7 (1982).
6. W. J. Youden, Accuracy of analytical procedures, *J. Assoc. Off. Agric. Chem.*, **45**, 169–73 (1962).
7. K. Doerffel, *Beurteilung von Analysenverfahren und -ergebnissen*, Springer-Verlag, Berlin, 1962.
8. K. Doerffel, Die statistische Auswertung von Analysenergebnissen. *Handbuch der Lebensmittelchemie Band II: Analytik der Lebensmittel. Teil 2: Nachweis und Bestimmung von Lebensmittel-Inhaltsstoffen*, Springer-Verlag, Berlin, 1194–1246, 1967.
9. W. J. Youden, *Statistical Techniques for Collaborative Tests*, Association of Official Analytical Chemists, Inc., Washington, DC, 1967.
10. BGA (Bundesgesundheitsamt) Amtliche Sammlung von Untersuchungsverfahren nach § 35 LMBG — Statistik (Mai 1983), *Planung und Auswertung von Ringversuchen*, Beuth-Verlag GmbH, Berlin, 1983.
11. H. Kaiser und H. Specker, Bewertung und Vergleich von Analysenverfahren, *Z. Analyt. Chem.*, **149**, 46–60 (1956).

PART II

RECOMMENDED METHODS

2
Vitamin A (Retinol and Retinyl Esters) in Food: HPLC Method

1. PURPOSE AND SCOPE

The method describes a procedure for the quantitative determination of total vitamin A content in food ready for consumption. Total vitamin A content comprises the naturally occurring vitamin A compounds in the constituents of the food as well as vitamin A which may have been added during its manufacture. The method may be applied to fresh and stored food and to food products intended for immediate consumption, e.g. milk, dried milk products, cheese, eggs, dried egg products, vegetables, potatoes, meat, fish and ready-cooked meals, beverages and butter, margarine and other oil and fat products. Quantitative determination of vitamin A content is possible down to 40 µg of retinol (133 IU of vitamin A) in 100 g of food.

2. DEFINITION

Vitamin A content is taken to mean the content determined by the method described here, excluding provitamin A carotenoids. It is given in µg of all-*trans*-retinol (vitamin A alcohol) per 100 g of sample; 1 µg of all-*trans*-retinol or 1·1467 µg retinyl acetate corresponds to 3·33 IUA; 1 µg retinyl acetate corresponds to 2·90 IUA.

3. BRIEF DESCRIPTION (PRINCIPLE OF THE METHOD)

After homogenisation and saponification of the material under investigation in a solution of potassium hydroxide in ethanol and water the retinol (vitamin A alcohol) released is totally extracted with a 1:1

mixture of petroleum ether and diethyl ether or alternatively with pure diethyl ether. Separation and determination of the retinol content are done with all or part of the extract by reversed-phase HPLC without separation of isomers. Measurement is carried out against an external vitamin A standard which has undergone the same procedure as the sample.

4. CHEMICALS

Remark: Unless otherwise specified, AR grade chemicals are to be used; water must be either distilled or of equivalent purity.

4.1. Potassium hydroxide, pellets, highest purity (content 85%), e.g. Merck.
4.2. Hydroquinone, highest purity, e.g. Fluka.
4.3. Sodium sulphide.(7–9 H_2O).
4.4. Retinyl acetate, e.g. type I Sigma grade, all-*trans*, synthetic, cryst., with approx. 2·9 million USP units/g, by Sigma (1 μg corresponds to approx. 2·9 IUA).
4.5. Peanut oil, free from vitamin A.
4.6. Ethanol, approx. 95% (for saponification).
4.7. Ethanol, absolute (for vitamin A standard solution).
4.8. Methanol, AR or spectrograde.
4.9. Diethyl ether, peroxide-free; stabilised with BHT.
4.10. Petroleum ether, boiling range 40–60 °C.
4.11. Nitrogen, oxygen-free, 99·9%(v/v).
4.12. Solutions.
  4.12.1. Potassium hydroxide solution, 50% (w/v) aqueous; dissolve 50 g of potassium hydroxide (4.1) in water (while cooling) to make 100 ml.
  4.12.2. Sodium sulphide solution; dissolve 12 g of sodium sulphide (4.3) in water to make 100 ml.
  4.12.3. Phenolphthalein solution, 2%; dissolve 2 g of phenolphthalein in ethanol (4.6) to make 100 ml.
  4.12.4. *Vitamin A standard solution.* Dissolve approx. 25 mg of retinyl acetate (4.4), weighed to within 0·01 mg, in absolute ethanol (4.7) to make 100 ml (=stock solution). Dilute 5·0 ml of stock solution (approx. 1·25 mg of retinyl acetate) to 100 ml with absolute ethanol (4.7) (=vitamin A standard solution with

approx. 12·5 μg of retinyl acetate/ml, corresponding to approx. 10·87 μg retinol or approx. 36·25 IUA/ml).

Note 1: Stock and standard solutions have to be prepared daily immediately before use. Don't keep them longer than 2 h at RT; then discard.

5. APPARATUS AND AUXILIARY EQUIPMENT

5.1. Rotary evaporator with water bath.
5.2. Water bath with magnetic stirrer, e.g. Colora.
5.3. Amber glass apparatus (not compulsory for working areas with UV protection).
   5.3.1. Ground-glass round-bottomed flasks, standard ground joint 29/32, 100, 250, 500 ml.
   5.3.2. Conical flask, standard ground joint 29/32, 250 ml.
   5.3.3. Graduated flasks 10, 50, 100 ml.
   5.3.4. Separating funnel, Squibb, conical, standard ground joint 29/32, with PTFE stopcock, 250, 500, 1000 ml.
5.4. Other glass apparatus (not amber glass).
   5.4.1. Allihn reflux condenser, jacket length 30 cm, male part standard ground joint 29/32.
   5.4.2. Adaptor for nitrogen feed, male part standard ground joint 29/32.
   5.4.3. Glass funnel, 18·5 cm dia.
5.5. Pleated filter for phase separation, 18·5 cm dia., e.g. Schleicher & Schüll, 597 Hy 1/2.
5.6. Apparatus for homogenising food samples (see 6.2): mixer, turbine mixer, mincer, sieves, etc.
5.7. HPLC apparatus, comprising:
   HPLC pump, e.g. Waters type 6000.
   20 and (or) 50 μl feed valve (loop).
   Columns for reversed-phase chromatography, e.g. (1) ready-made column, C_{18}, Bondapak 10 μm; 30 × 0·39 cm I.D. by Waters or similar (also with 5 μm particle size, 25 cm column length) (self-filling); (2) ready-made column, C_8, Supelcosil by Supelco.
   Photometric HPLC detector for UV, e.g. Waters.
   Integrator.
   Recorder.
5.8. Membrane filter, 0·45 μm, e.g. Millipore or Sartorius.

6. SAMPLE

6.1. Sampling

The proportional composition of the sample to be taken must be representative of the material under investigation (the food as a whole). In any case, the initial sample must be fairly large (about 5–10 times the weighed portion in cases where the composition of the food is not uniform).

6.2. Sampling Method and Preparation of the Sample

The sample must be comminuted as finely as possible (mixer, chopper, mincer, sieves, etc.) and homogenised. After preparation it is in powder, pulp, suspension, solution or molten (oils, fats) form. It must be analysed immediately and quickly.

6.2.1. *Weighed portion.* The weighed portion should be limited to 50 g max. because of the equipment used and should contain at least 30 µg (100 IU) of vitamin A. Examples: cheese 10 g, milk 40 ml, fats and oils 1–20 g. When the vitamin A content in extremely low, two or more maximum weighed portions must be prepared separately but in parallel and the extracts subsequently combined.

> *Note 2:* Combining several different preparations can lead to overloading of the sample test solution (7.4) with impurities which have to be removed by an appropriate clean-up operation before HPLC is used (e.g. by conventional column chromatography on deactivated Al_2O_3 or by preparative HPLC). This applies chiefly to animal and plant material.

7. PROCEDURE

7.1. Remark

Vitamin A is sensitive to UV radiation and oxygen. The work is done under exclusion of UV light (amber glass apparatus, UV-absorbent or aluminium foil) and oxygen (nitrogen flushing). In particular, the air over the liquid must be replaced by nitrogen (4.11). Solutions must be evaporated at reduced pressure by means of a rotary evaporator at a temperature below 40 °C. For the effects of solvents on retinyl esters see Ref. 10.16.

7.2. Saponification

Once the material to be examined has been prepared and weighed into a 250 or 500 ml ground-glass round-bottomed flask (6.2; 6.2.1), ethanol (4.6) (volume in millilitres is three–four times the weighed portion in grams, but at least 30 ml), 50% potassium hydroxide solution (4.12.1) (volume as weighed portion), approx. 100 mg of hydroquinone (4.2) and—if the material contains minerals—2 ml of sodium sulphide solution (4.12.2) are successively swirled into it.

The reaction mixture is kept boiling under reflux on the water bath at 80 °C for 30 min, while stirring continuously (magnetic stirrer) or swirling several times in a slow nitrogen stream (4.11) (introduced via the reflux condenser). After saponification, the condenser is rinsed with about 20 ml of water and the contents of the flask are cooled in running water.

Note 3: Saponification is indispensable for determining the total vitamin A content, owing to the various natural retinyl esters present and their binding to the substrate. Saponification carried out overnight at room temperature gives rise to considerable losses, because retinol which has already been released is then exposed to the action of the alkali for too long. In the same way, enzymatic digestion is associated with the destruction of released retinol.

7.3. Extraction

The contents of the flask are transferred without loss into a 500 (1000) ml separating funnel with enough water to bring the ethanol/water ratio to at least 1:1 and, in the case of small weighed portions (1–5 g), to 1:2 and 1: > 2. Total extraction is carried out several times with diethyl ether (4.9) (e.g. 4 × 150 ml) or a 1:1 mixture of petroleum ether (4.10) and diethyl ether (4.9) (e.g. 2 × 500 ml). The number and size of the extraction steps must be so matched to the behaviour of the solution to be extracted that the product of the next extraction will be free of vitamin A (checked by HPLC).

Note 4: In the case of products rich in fats, proteins and starch, intensified multiple extraction may be necessary. Single step extraction with larger volumes, e.g. in the case of low weighed portions, is unsuitable for analysing food for vitamin A because it is not exhaustive.

In a 1000 ml separating funnel the combined extracts are carefully washed several times with water saturated with diethyl ether (4.9), either as a whole or in aliquots, in order to remove alkali (avoiding emulsification), until the washing water (4 × 100 ml of water is generally sufficient) remains colourless when phenolphthalein (4.12.3) is added.

Note 5: Emulsions can occasionally be broken by adding ethanol, and in alkali extraction by adding common salt to the washing water (first wash).

For drying, the extract is filtered through a dry pleated phase-separation filter (5.5) and the separating funnel and filter are carefully flushed with approx. 50 ml of extracting agent.

Note 6: If this type of filter is used it is not necessary to dry the extract with anhydrous sodium sulphate.

For further processing, the entire extract or, when there is a high vitamin A content, an aliquot part—obtained by placing the entire extract in a suitable graduated flask, filling up to the mark then taking the aliquot from this solution—is evaporated almost to dryness in portions in a 250 ml conical flask (5.3.2) under vacuum at a maximum bath temperature of 40 °C in the rotary evaporator.

Note 7: If the saponification solution is of a suitable consistency the aliquot can be taken from it: the solution is placed in a graduated flask and made up to the mark or to a given weight with water, and aliquot volumes or weights are then taken (saving of extraction agent).

If the concentrate is still turbid (water!), a few millilitres of ethanol (4.7) are added and the solvent is evaporated again, the water being distilled off azeotropically. Any residual solvent is blown off with nitrogen (4.11) and the conical flask is sealed.

7.4. *Sample Test Solution for HPLC*

The residue in the flask (7.3) is immediately dissolved in a quantity of methanol (4.8) such that the resulting test solution contains approximately 1·5–4·5 µg of retinol, corresponding to about 5–15 IUA per ml.

7.5. *Standard Test Solution for HPLC*

A vitamin A standard solution is treated in the same manner as the sample in order to correct the vitamin A losses which are unavoidable when the sample is being analysed (indirect recovery).

Note 8: Unavoidable losses occur during washing of the extract, evaporation, isomerisation, etc.

A quantity (5·0 ml) of vitamin A standard solution (4.12.4) (approx. 62·5 µg retinyl acetate), in a 100 ml round-bottomed flask with 25 ml of ethanol (4.6), 10 ml of 50% potassium hydroxide (4.12.1), approx. 0·5 ml of peanut oil (4.5) and 100 mg of hydroquinone (4.2) is kept boiling under reflux on the water bath at 80 °C for 30 min in a nitrogen atmosphere (4.11).

> *Note 9:* Experience shows that peanut oil has a protective effect on released retinol during saponification, extraction and evaporation to dryness.

The condenser is rinsed with 20 ml of water and the flask contents are cooled and transferred without loss into a 250 ml separating funnel while rinsing with 20–30 ml of water and with some of the extracting solvent. Extraction is carried out twice with 100 ml of diethyl ether (4.9) or a 1:1 mixture of petroleum ether (4.10) and diethyl ether (4.9), depending on the sample. The extract is washed alkali-free, dried and evaporated almost to dryness.

The residue is dissolved in methanol (4.8) to 10·0 ml (= standard test solution with approx. 5·4 µg retinol or 18 IUA/ml); either 20 or 50 µl of this solution, i.e. approx. 109 ng retinol (0·36 IUA) or 272 ng retinol (0·9 IUA), respectively, is then injected onto the HPLC column. The peak height obtained (H_S) and the concentration of the vitamin A standard test solution (C_S) are to be used as recovery correction in the equation for the final calculation (8.1).

7.6. HPLC

Complete separation and measurement of retinol are done by means of reversed-phase HPLC, using C_8 or C_{18} RP columns. The total vitamin A of the sample is determined as total retinol without isomer separation.

> *Note 10:* If it is desired to determine separately the retinol isomers which are sometimes found in food, straight phase HPLC must be used.

The sample test solution and the standard test solution are chromatographed simultaneously.

Chromatographic conditions

Stationary phase: C_8 or C_{18} reversed-phase column packing (5.7).

Mobile phase: methanol/water in various ratios (e.g. 85:15, 90:10, 95:5, 98:2; to be determined experimentally in each case).

Volume to be applied for sample test solution (7.4): 20 or 50 µl (loop) as required, with retinol quantities of, e.g., 30–90 ng or 75–220 ng, corresponding to 0·10–0·27 and 0·25–0·73 IU of vitamin A respectively.

Volume to be applied for standard test solution (7.5): 20 or 50 µl as required, with retinol quantities of 109 or 272 ng, corresponding to 0·36 and 0·9 IU of vitamin A respectively.

Sample test solution and standard test solution are to be ultrafiltered before injection (5.8).
Detection: 325 nm.
Peak evaluation: By peak height or area (height is preferable) with integrator or manually.
Flow rate: 1 ml/min.
Sensitivity: from 0·05 to 0·2 AUFS as required.
Retention time for retinol: to be determined.
Internal standard (if required): retinyl acetate (to be added for extraction after saponification); used only for determining retention time.

8. EVALUATION

8.1. *Calculation*

The content of total vitamin A in the material being analysed is calculated from the peak heights (or areas) for retinol of the microlitres (µl) of sample test solution and standard test solution injected (same microlitre quantities!); it is already recovery-corrected:

$$\text{(a)} \quad \mu g \text{ retinol}/100 \, g = \frac{H_P C_S M \times 100}{H_S E a}$$

where: H_P = peak height for retinol in sample test solution (mm); H_S = peak height for retinol in standard test solution (mm); C_S = concentration of standard test solution (µg retinol/ml); M = quantity of methanol for preparation of standard test solution (ml); E = weighed portion of sample (g); a = size of aliquot part of extract (or of saponification solution) (as a fraction). In the case of area evaluation, F_P (peak area for retinol in sample test solution) and F_S (peak area for retinol in standard test solution) are to be substituted accordingly.

$$\text{(b)} \quad \text{IUA}/100 \, g = (\mu g \text{ retinol}/100 \, g) \times 3 \cdot 33$$

8.1.1. *Range of linearity.* Within the range of retinol quantities (30–270 ng) applied in the various injection volumes of sample and standard test solution, there is linearity in the indication range. The values given in the literature are in the range of 20–320 ng.

8.1.2. *Detection limit.* Quantitative detection is possible down to 20 ng.

8.2. Reliability of the Method

8.2.1. *Recovery.* The recovery rate for this method has not been experimentally determined. In similar methods for the determination of vitamin A, the values obtained were 94–96% when retinyl acetate was added to the starting material (assuming complete extraction). According to the literature, the recovery rates for pure, matrix-free, analysed solutions of vitamin A are likewise around 96%.

8.2.2. *Repeatability.* This has not been determined for this method. For similar HPLC–RP methods the literature gives the following relative standard deviations for sample or standard: $S_{rel}\% = \pm 1.0$ (Ref. 10.5); ± 2.0 and 2.6 (Ref. 10.12); ± 3.1–3.5 (Ref. 10.10); ± 3.6–3.9 (Ref. 10.11) and ± 4.0 (Ref. 10.8).

8.2.3. *Comparability.* This has not been determined for this method (no syndicate tests).

8.2.4. *Comparison with other vitamin A determination methods.* According to the literature and to private communications the HPLC findings are all up to 4–5% lower than corresponding Carr–Price findings. The differences as compared with the results of UV determination of extracts are even greater (see also Ref. 10.7).

HPLC is used for determining the *intact* total vitamin A.

9. ANALYSIS REPORT

The result of the determination is to be given with a reference to this method. Operations not mentioned in the described method have to be indicated.

10. REFERENCES

The determination method described was worked out on the basis of the following literature on HPLC reversed-phase vitamin A determinations, etc., and of private communications.

10.1. R. A. Wiggins, E. S. Zai and I. Lumley, *Chromatographic Sciences, Applied Liquid Chromatog.*, 4 (485508), **20**, 327–41 (1982).

10.2. P. A. Jackson, *Food*, 12–18, (Dec. 1982).
10.3. B. Stancher and F. Zonta, *J. Chromatog.*, **238**, 217–25 (1982).
10.4. M. I. R. M. Santoro, J. F. Magalhaes and E. R. M. Hackmann, *J. Assoc. Off. Anal. Chem.*, **65** (3), 619–23 (1982).
10.5. O. Bohman, K.-A. Engdahl and H. Johnsson, *J. Chem. Education*, **59** (3), 251–2 (1982).
10.6. H. Rückemann, *Z. Lebensm. Unters. Forsch.*, **173**, 113–16 (1981).
10.7. J. Blattná and H. Pařizková, *Mitt. Gebiete Lebensm. Hyg.*, **73**, 166–73 (1982).
10.8. Mai-Huong Bui-Nguyen and B. Blanc, *Experientia*, **36**, 374–5 (1980).
10.9. B. S. Kamel and M. Bueno, *Lebensm, -Wiss, u. Technol.*, **13**, 134–7 (1980).
10.10. F. Zonta, B. Stancher and C. Calzolari, *Rassegna Chimica*, **32** (6), 301–8 (1980).
10.11. D. C. Egberg, J. C. Heroff and R. H. Potter, *J. Agric. Food Chem.*, **25** (5), 1127–32 (1977).
10.12. D. B. Dennison and J. R. Kirk, *J. Food Science*, **42** (5), 1376–9 (1977).
10.13. T. van de Weerdhof, M. L. Wiersum and H. Reissenweber, *J. Chromatog.*, **83**, 455–60 (1973).
10.14. D. Scuffam, Laboratory of the Government Chemist, Department of Industry, London, 1981. Private communication.
10.15. U. Manz, F. Hoffman-La Roche & Co. AG, Vitamins and Nutrition Research Department, Basle, Switzerland, 1981. Private communication.
10.16. M. C. Mulry, R. H. Schmidt and J. R. Kirk, *J. Assoc. Off. Anal. Chem.*, **66** (3), 746–50 (1983).

3

Carotene in Foodstuffs

1. PURPOSE AND SCOPE

The method describes three procedures for the quantitative determination of the total carotene in foodstuffs, that is the total of all-*trans* carotenes (α-, β-, γ-, δ- and ε-) and their stereoisomers of natural origin. β-Carotene, added during the production of foodstuffs, is included in this assay.

It is necessary to use three different procedures depending on the material under investigation and optimised for the purpose required.

Procedure A: *Determination of total carotene in complex foodstuffs*
Procedure B: *Determination of total natural carotene in fruits, vegetables and unaltered plant materials*
Procedure C: *Determination of total carotene in beverages*

Carotene contents of 0·1 mg/100 g (or ml) of sample and above can be determined quantitatively.

2. DEFINITION

Total carotene is the quantity of carotene determined by the method described below; it is calculated as β-carotene and given in mg carotene/100 g (or ml) of sample. No attempt is made to separate carotene homologues and isomers in this method; the small amount of α-carotene and isomers occurring in foodstuffs means that separation is not necessary.

PROCEDURE A: DETERMINATION OF TOTAL CAROTENE IN COMPLEX FOODSTUFFS

1. SCOPE

Procedure A can be used for fresh and preserved foodstuffs of complex composition for direct consumption such as complete meals, ready-to-serve dishes, etc. Vegetables, fruits, offals and milk products are considered to be carotene-containing components of these foodstuffs. The determination of carotene in original plant materials and beverages is not covered by this procedure.

2. DEFINITION

See Section 2, p. 33.

3. BRIEF DESCRIPTION (PRINCIPLE OF THE METHOD)

Following alkaline saponification of the sample, exhaustive extraction with diethyl ether and taking up in hexane, the total carotene is chromatographically separated on a column of deactivated aluminium oxide. The eluate contains the total carotene which is determined by extinction measurements using a spectrophotometer.

4. CHEMICALS

Remark: Unless otherwise specified, AR grade chemicals are to be used. Water must be either distilled or of equivalent purity.

4.1. Potassium hydroxide, pellets, highest purity (content 85%), e.g. Merck.
4.2. Hydroquinone, highest purity, e.g. Fluka.
4.3. Sea sand, cleaned with acid, glowed, e.g. Merck.
4.4. Aluminium oxide 90 active, neutral (activity level I), e.g. Merck.
    4.4.1. Aluminium oxide, deactivated: 100 g of aluminium oxide (4.4) is poured into a 200 ml conical flask with 12 ml of water; shake until all lumps have disappeared and allow to stand for 2 h.

Note 1: The deactivated aluminium oxide represents the activity level IV–V. Since the activity changes after long periods of storage the required quantities should be made up fresh at least 2 h before use.

4.5. pH paper, universal indicator paper pH 1–11, e.g. Macherey–Nagel.
4.6. *n*-Hexane, puriss. (analytical grade), e.g. Merck.
4.7. Diethyl ether AR, stabilised with BHT, e.g. Merck.
4.8. Petroleum ether, boiling range up to approximately 40 °C, e.g. Merck.
4.9. Ethanol, absolute, e.g. Merck.
4.10. Nitrogen, oxygen-free, 99·9% (v/v).
4.11. Potassium hydroxide solution, 50% (w/v) aqueous; dissolve 50 g KOH (4.1) in water to make 100 ml (cool!).

5. APPARATUS AND AUXILIARY EQUIPMENT

5.1. Standard basic laboratory equipment.
5.2. Mixing apparatus, e.g. Waring blender; Polytron rod.
5.3. Coffee grinder or similar.
5.4. Thermostatic water bath.
5.5. Separating funnel, Squibb conical; 200, 250, 500 ml, with PTFE stopcock.
5.6. Reflux condenser, NS 29.
5.7. Gas inlet tube suitable for (5.6).
5.8. Graduated ground-glass shaking cylinder, 500 ml.
5.9. Shaking machine.
5.10. Magnetic stirrer.
5.11. Rotary evaporator.
5.12. Chromatography tube of Duran 50, diameter 2 cm with fritted glass bottoms (G_1, P_1) and stopcock or Allihn tube.
5.13. Spectrophotometer, commercially available.
5.14. Two glass cuvettes, preferably 1 cm layer thickness.

6. SAMPLE

6.1. *Sampling*

The sample must be of proportional composition representative of the material (foodstuff) to be analysed. It must in any case be taken from a large initial quantity (about 500 g).

6.2. Sampling Method and Preparation of the Sample

Carefully taken average samples of at least 500 g should be thoroughly mixed and reduced in size (mincing knife, mixing apparatus, Polytron rod, grinder, sieve) (5.2; 5.3) and homogenised. All the work should be carried out in subdued light or in amber glass apparatus, if necessary under nitrogen (4.10).

6.2.1. Weighed portion.
The size of the weighed portion of the foodstuff prepared according to method 6.2 will depend on the amount of carotene expected and should be a maximum of 30 g.

7. PROCEDURE

7.1. Remark
The method is best carried out without major interruption as a true double determination (two weighed portions); direct sunlight should be avoided.

7.2. Saponification
The sample is weighed out into a 250 ml ground-glass round flask and treated in succession with 60 ml of ethanol (4.9), 10 ml of potassium hydroxide solution (4.11), 10 ml of petroleum ether (4.8) and approximately 20 mg of hydroquinone (4.2) or more as long as the mix ratio is maintained, shaken or stirred for 30 min (magnetic stirrer (5.10)) on a water bath at a temperature of approximately 60 °C and kept boiling under reflux (5.6) and nitrogen feed (5.7). At the end of the saponification the reflux condenser is washed out with approximately 20 ml of water, and the contents of the flask cooled to room temperature (= saponification solution).

> *Note 2:* The addition of petroleum ether is effected in order to make the saponification less harsh (temperature reduction, protective effect on the released carotene).

7.3. Extraction
The carotene can be isolated from the saponification solution (7.2) by single or multiple extraction depending on the type of sample. The final extract should be colourless.

7.3.1 *Single extraction.* The saponification solution (7.2) is transferred without loss with a maximum of 100 ml of water in a 500 ml graduated ground-glass shaking cylinder (5.8). After addition of approximately 250 ml of diethyl ether (4.7), the saponification flask being rinsed, the cylinder is shaken intensively for 5 min on a shaking machine (5.9). After complete separation of the two phases, the volume of the ether phase is determined (for subsequent calculation).

An aliquot part of the ether phase (quantity according to colour intensity but at least 25 ml) is washed in a suitable Squibb separating funnel (5.5) with ether-saturated water to make it alkali free (pH paper (4.5) or phenolphthalein control). The neutral ether extract is evaporated under vacuum in a suitable round flask at approximately 40 °C in the rotary evaporator (5.11).

The remaining traces of water are removed by the addition and evaporation of ethanol (4.9). The dry residue is again evaporated with a certain amount of *n*-hexane (4.6) to remove traces of ethanol and finally taken up in 3·5 ml of *n*-hexane (4.6) (= hexane solution of unsaponifiables).

7.3.2. *Multiple extraction.* If necessary, extraction is carried out three times after the addition of a maximum of 100 ml of water with 100 ml of diethyl ether (4.7) in a 500 ml Squibb separating funnel (5.5). The combined ether extracts or an aliquot part thereof are then further treated as in section 7.3.1.

7.4. *Separation of Carotene by Column Chromatography*
Using a chromatography tube or an Allihn tube (5.12), the fritted plate is covered with a layer of sea sand of approximately 0·5 cm thickness (4.3); a suspension of deactivated aluminium oxide (4.4.1) in *n*-hexane (4.6) is poured slowly in and the sedimenting particles are distributed by lightly shaking the column.

The aluminium oxide column, thus prepared, should be approximately 8–9 cm high and permanently covered with *n*-hexane. The hexane solution of unsaponifiables (7.3.1; 7.3.2) is transferred by means of a dropping pipette with subsequent washing with small portions of *n*-hexane (4.6) on to the top of the column and the yellow-orange carotene which moves in front is subsequently eluted with *n*-hexane (4.6) into a suitable graduated flask, e.g. 100 ml. The process is terminated when the last fraction of eluate is shown spectrophotometrically to contain no more carotene colouring. The graduated flask is filled up to the mark with *n*-hexane (4.6) (= test solution).

Note 3: The column can take up to 200 μg total carotene without any deleterious effect on the separating capacity. Accompanying carotenoids remain on the column: the xanthophylls at the top of the column, the lycopin in the middle.

7.5. Spectrophotometric Measurement

7.5.1. Standard. There is no requirement to use a standard since the calculations can be carried out with the experimentally determined and generally valid extinction of the β-carotene in *n*-hexane (4.6):

$E_{1\,cm}^{1\%}$ = Extinction of a 1% β-carotene solution in *n*-hexane (in a 1 cm cuvette at the maximum near 450 nm) = 2590.

7.5.2. Measurement. Using *n*-hexane (4.6) as a blank solution, the extinction of the test solution (7.4), preferably in a 1 cm cuvette (5.14) is measured in the normal way in the spectrophotometer (5.13).

8. EVALUATION

8.1. Calculation

The total carotene content is calculated as β-carotene on the basis of its specific extinction in *n*-hexane (7.5.1) according to the formula:

$$\text{Total carotene (mg/100 g sample)} = \frac{\text{Ext} \times V_M \times 100^a \times 1000}{2590 \times 100^b \times A \times d}$$

where: Ext = extinction of the test solution (7.4) (scale divisions); $2590 = E_{1\,cm}^{1\%}$ of β-carotene in *n*-hexane (scale divisions); V_M = volume of the test solution (ml); d = layer thickness of the cuvette (cm); A = weighed portion of the product or proportion of the weighed portion in a checked aliquot part of the ether extract (7.3.1; 7.3.2) (g); $A = V_A E/V_E$, where, E = weighed portion (g), V_E = volume of ether extract (ml) and V_A = volume of aliquot part of ether extract (ml); 100^a = conversion to 100 g of sample; 100^b = conversion to 1 ml of test solution; 1000 = conversion to mg of carotene.

8.1.1. Range of linearity. The extinction is directly linear and proportional to the concentration in the range 100–5000 ng/ml test solution.

8.1.2. Detection limit. An amount of 100 ng of β-carotene/ml of test solution can be quantitatively determined.

8.2. Reliability of the method

8.2.1. Recovery. No obligatory recovery determination is required in this method. Experience shows the average recovery rate to be 95%.

8.2.2. Repeatability. This has been determined for the DGF-standard method F-II 2a (75), which is in principle the same, as follows (see Ref. 10.3):

Lab. 1: standard deviation ± 0.050; rel. standard deviation $\pm 2.91\%$;
Lab. 2: standard deviation ± 0.021; rel. standard deviation $\pm 1.24\%$
(for $N = 3$ and 1·7 mg carotene/100 g peanut oil).

8.2.3. Comparability. No details are available (no syndicate tests).

8.2.4. Comparison with other carotene determination methods. No comparisons are available.

PROCEDURE B: DETERMINATION OF TOTAL NATURAL CAROTENE IN FRUITS, VEGETABLES AND UNALTERED PLANT MATERIALS

1. SCOPE

Procedure B can be used for fresh and preserved vegetable products which have been prepared for direct consumption; it is not applicable to complete meals, ready-to-serve dishes and preparations of vegetables which contain other food components, nor to beverages.

2. DEFINITION

See Section 2, p. 33.

3. BRIEF DESCRIPTION (PRINCIPLE OF THE METHOD)

Dried plant material (e.g. dried vegetables) is extracted exhaustively with *n*-hexane/acetone mixture (or petroleum ether/acetone mixture) at room temperature and the extract saponified without heating with a methanol potassium hydroxide solution to break down the chlorophylls. Fresh vegetable material with more than 15% water content is treated in the same way but with acetone only. The extract, evaporated to dryness and taken up in *n*-hexane or petroleum ether, is chromatographed on partly deactivated aluminium oxide and thus the carotene fraction is separated from the xanthophyll fraction. The eluate contains the total carotene which is determined by extinction measurements with the spectrophotometer.

4. CHEMICALS

Remark: Unless otherwise specified, AR grade chemicals are to be used. Water must be either distilled or of equivalent purity.

4.1. Potassium hydroxide, pellets, highest purity (content 85%), e.g. Merck.
4.2. Sodium sulphate, free from water, fine powder, extremely pure, e.g. Merck.
4.3. Aluminium oxide 90 active, neutral (activity level I), e.g. Merck.
 4.3.1. Aluminium oxide, deactivated: 100 g of aluminium oxide (4.3) is poured into a 200 ml conical flask with 12 ml of water; shake until all lumps have disappeared, and allow to stand for 2 h.

Note 4: Deactivated aluminium oxide represents the activity level IV–V. Since the activity changes after long periods of storage the required quantities should be made up fresh at least 2 h before use.

4.4. *n*-Hexane, puriss., e.g. Merck or puriss. p.a., e.g. Fluka.
4.5. Petroleum ether, boiling range 40–60 C, e.g. Merck.
4.6. Petroleum ether, puriss., boiling range 50–70 °C, Merck.
4.7. Acetone, e.g. Merck or pure 99%, e.g. Fluka.
4.8. Methanol, e.g. Merck or pure 99%, e.g. Fluka.
4.9. Ethanol, absolute, e.g. Merck.
4.10. Nitrogen, oxygen-free, 99·9% (v/v).
4.11. *n*-Hexane (4.4) (or petroleum ether (4.6))–acetone (4.7) mixture, 70:30 (v/v).
4.12. Potassium hydroxide, 40% (w/v) in methanol; 40 g KOH (4.1) are dissolved in methanol (4.8) to 100 ml.

5. APPARATUS AND AUXILIARY EQUIPMENT

5.1. Standard basic laboratory equipment including conical flasks, separating funnels (Squibb), mortars, pipettes, graduated flasks, graduated cylinders with ground stoppers.
5.2. Water bath.
5.3. Rotary evaporator.
5.4. Mixing apparatus, e.g. mincing knife, sieves, Polytron rod.
5.5. Homogeniser, e.g. Waring blender.
5.6. Chromatography tube: 30 cm high, 10–15 mm diameter, or Allihn tube with G_1 or P_1 fritted glass bottoms.
5.7. Spectrophotometer, commercially available.
5.8. Two glass cuvettes, preferably 1 cm layer thickness.

6. SAMPLE

6.1. *Sampling*
The sample must be of proportional composition representative of the vegetable material to be analysed. It must in any case be taken from a large initial quantity (about 500 g). If the substance to be analysed is to be stored or preserved this must be in a refrigerator. Fresh plant material must be stored in the deep-frozen state.

6.2. *Sampling Method and Preparation of the Sample*
Carefully taken average samples weighing at least 500 g should be thoroughly mixed and reduced in size (mincing knife, mixing apparatus, 0·5 mm sieve, Polytron rod) (5.4) and homogenised (mixer; Waring blender) (5.5). All the work should be carried out in subdued light or in amber glass apparatus, if necessary under nitrogen (4.10).

6.2.1. *Weighed portion.* The size of the weighed portion of the material prepared according to method 6.2 will depend on the amount of carotene expected. For dried vegetable material it will amount to 1·5–3·0 g; for fresh material with a high water content it will be approximately 10 g.

7. PROCEDURE

7.1. *Remark*
The method is best carried out without major interruption as a true

double determination (two weighed portions) because of the sensitivity of carotenes to light and oxidation. In any case direct sunlight should be avoided.

7.2. Disintegration, Saponification and Extraction

7.2.1. For dried plant material (up to 15% water content).
A quantity (1·5–3·0 g) of finely-reduced sample material is poured into an amber 100 ml graduated flask with 30 ml hexane (petroleum ether)—acetone mixture (4.11) and allowed to stand for extraction overnight under nitrogen (4.10).

> *Note 5:* Once petroleum ether has been used, it must continue to be used for all subsequent operations (instead of hexane). Two petroleum ether fractions must be used!

One hour before the chromatographic operation, 2 ml of the potassium hydroxide solution in methanol (4.12) is added. The sample is shaken vigorously and then allowed to stand for 30 min at room temperature. Two millilitres of water are then added; the sample is again shaken and allowed to settle. The sample is then made up to 100 ml with *n*-hexane (4.4) or petroleum ether (4.6); the solution is thoroughly mixed and allowed to settle. An aliquot is carefully evaporated almost to dryness under vacuum in a rotary evaporator (5.3) at a water bath temperature of 50 °C, reduced again with a certain amount of ethanol (4.9) to remove any residual water and finally taken up in 3–5 ml *n*-hexane (4.4) or petroleum ether (4.5) (=extract concentrate).

7.2.2. For fresh plant material.
A quantity (10 g) of finely-cut plant material is reduced, homogenised and extracted in succession three times with 50 ml acetone (4.7) in a homogeniser (5.5), the safety instructions being carefully observed. The acetone extracts are separated by means of a G_3 suction filter, combined and made up to the mark with acetone (4.7) in a 200 ml graduated flask.

> *Note 6:* Deep-freezing of the acetone extract involves carotene losses and should therefore be avoided where possible.

A quantity (20 ml) of the solution is poured into a separating funnel and 0·5 ml potassium hydroxide solution in methanol (4.12) is added. This is shaken vigorously, allowed to stand for 30 min, and then 30–40 ml *n*-hexane (4.4) or petroleum ether (4.6) are added and it is shaken vigorously. To remove the acetone and the potassium hydroxide, the solution is washed three times with 10 ml water and the extract treated as described in section 7.2.1 (=extract concentrate).

7.3. Chromatography

Using a chromatography tube or an Allihn tube (5.6) a suspension of deactivated aluminium oxide (4.3.1) in n-hexane (4.4) or petroleum ether (4.5) is poured slowly in and the sedimenting particles are distributed by lightly shaking the column.

The aluminium oxide column, thus prepared, should be approximately 8–15 cm high and permanently covered with the solution medium (4.4 or 4.5); it may be covered with an approximately 20 mm thick layer of sodium sulphate (4.2). The extract concentrate (7.2.1; 7.2.2) is completely transferred by means of a dropping pipette into the column and this is then rinsed with several small portions of n-hexane (4.4) or petroleum ether (4.5).

The yellow-orange carotene which moves in front is subsequently eluted with n-hexane (4.4) or petroleum ether (4.5), without pressure, in a suitable graduated flask (e.g. 100 ml). The process is terminated when the last fraction of eluate is shown spectrophotometrically to contain no more carotene colouring. The graduated flask is filled up to the mark with n-hexane (4.4) or petroleum ether (4.5) (=test solution).

Note 7: In general, the loss of colour of the eluate indicates the end of carotene elution.

The xanthophylls remain at the top of the column.

7.4. Spectrophotometric Measurement

7.4.1. *Standard.* There is no requirement to use a standard since the calculations can be carried out with the experimentally determined and generally valid extinction of β-carotene in n-hexane (4.4) and petroleum ether (4.5):

$E_{1\,cm}^{1\%}$ = Extinction of a 1% β-carotene solution in n-hexane (in a 1 cm cuvette at the maximum near 450 nm) = 2590 and in petroleum ether (boiling range 40–60 °C) = 2600

7.4.2. *Measurement.* Using n-hexane (4.4) or petroleum ether (4.5) as a blank solution, preferably in a 1 cm cuvette (5.8), the extinction of the test solution (7.3) is measured in the normal way in the spectrophotometer (5.7).

8. EVALUATION

8.1. Calculation

The total carotene content is calculated as β-carotene on the basis of its specific extinction in n-hexane (4.4) or petroleum ether (4.5) (7.4.1) according to the formula:

$$\text{Total carotene (mg/100 g sample)} = \frac{\text{Ext} \times 100^a \times 1000 \times V_M \times V_E}{2590[2600] \times 100^b \times V_A \times E \times d}$$

where: Ext = extinction of the test solution (7.3) (scale divisions); $2590 = E^{1\%}_{1\,\text{cm}}$ of β-carotene in n-hexane (scale divisions); $2600 = E^{1\%}_{1\,\text{cm}}$ of β-carotene in 40–60 °C petroleum ether (scale divisions); V_M = volume of the test solution (7.3) (ml); V_E = volume of hexane (petroleum ether) or acetone extract (7.2.1 or 7.2.2) (ml); V_A = volume of aliquot of the extracts (ml); E = weighed portion (g); d = layer thickness of the cuvette (cm); 100^a = conversion to 100 g sample; 100^b = conversion to 1 ml test solution; 1000 = conversion to mg of carotene.

8.1.1. Range of linearity. The extinction is directly linear and proportional to the concentration in the range 100–5000 ng/ml test solution.

8.1.2. Detection limit. An amount of 100 ng of β-carotene/ml test solution can be quantitatively determined.

8.2. Reliability of the Method

8.2.1. Recovery. No obligatory recovery determination is required in this method since the isolation of carotenes from vegetable tissue (cells) is not analytically comparable with the separation of added free β-carotene.

8.2.2. Repeatability. No details are available. The difference between the two parallel determinations must not exceed 8%.

8.2.3. Comparability. No details are available (no syndicate tests).

8.2.4. Comparison with other carotene determination methods. The nondestructive extraction of carotene from plant material described here always provided the highest values in comparison with the use of alkaline disintegration (Roche Control Department, Grenzach; 1970–79). n-Hexane was used throughout as solvent.

PROCEDURE C: DETERMINATION OF TOTAL CAROTENE IN BEVERAGES

1. SCOPE

Procedure C can be used for beverages such as fruit preparations, vitamin juices, fruit milk, fruit buttermilk, vitamin drinks, etc.

> *Note 8:* Since the material under investigation is not saponified, it is possible that carotene bound to the matrix will not be detected. In commercially available beverages this proportion should not be significant.

2. DEFINITION

See Section 2, p. 33.

3. BRIEF DESCRIPTION (PRINCIPLE OF THE METHOD)

The beverage sample is extracted in a single course with chloroform, the extract (aliquot) chromatographed on deactivated aluminium oxide and the extinction of the carotene fraction measured sprectrophotometrically in n-hexane at the maximum near 450 nm.

4. CHEMICALS

> *Remark:* Unless otherwise specified, AR grade chemicals are to be used. Water must be either distilled or of equivalent purity. When using chloroform the relevant safety precautions are to be taken.

4.1. Sea sand, cleaned with acid, glowed, e.g. Merck.
4.2. Sodium sulphate, free from water, fine powder, extremely pure, e.g. Merck.

4.3. Aluminium oxide 90 active, neutral (activity level I), e.g. Merck.

4.3.1. Aluminium oxide, deactivated: 100 g Al_2O_3 (4.3) is poured into a 200 ml conical flask with 12 ml of water; shake until all lumps have disappeared and allow to stand for 2 h.

Note 9: Deactivated aluminium oxide represents the activity level IV–V. Since the activity changes after long periods of storage, the required quantities should be made up fresh at least 2 h before use.

4.4. n-Hexane, puriss., e.g. Merck.
4.5. Ethanol, absolute, e.g. Merck.
4.6. Chloroform, stabilised with ethanol, e.g. Merck.

5. APPARATUS AND AUXILIARY EQUIPMENT

5.1. Standard basic laboratory equipment.
5.2. Ground-glass shaking cylinder or ground-glass round flask furnished with indentations, 500 ml.
5.3. Shaking machine, commercially available, e.g. type W-3 Hormuth–Vetter.
5.4. Laboratory centrifuge.
5.5. Rotary evaporator.
5.6. Chromatography tube of Duran 50, diameter 2 cm, with fritted glass bottoms (G_1, P_1) and stopcock or Allihn tube.
5.7. Spectrophotometer, commercially available.
5.8. Two glass cuvettes, preferably 1 cm layer thickness.

6. SAMPLE

6.1. *Sampling*
The sample must be of proportional composition representative of the material (beverage) to be analysed.

6.2. *Sampling Method and Preparation of the Sample*
Beverages which are not clear or which contain some sediment must be shaken well before sampling.

6.2.1. *Weighed portion.* Depending on the type of beverage, a weighed portion of 50 ml or 50 g will be taken.

7. PROCEDURE

7.1. Remark
The method is best carried out rapidly as a true double determination (two weighed portions); direct sunlight should be excluded.

7.2. Extraction
50 ml or 50 g of the beverage, weighed to an accuracy of 0·1 g, are poured into a 500 ml ground-glass round flask with indentations (5.2); 200–400 ml of chloroform (4.6), accurately measured, are added, and the mixture is then shaken strongly for 30 min in the shaking machine (5.3). The chloroform/water emulsion which results is centrifuged on a laboratory centrifuge (5.4), the aqueous (upper) phase sucked away and the remaining chloroform extract filtered through a cotton wool plug with a layer of sodium sulphate (4.2) (=sample extract). An aliquot part of the sample extract is evaporated under vacuum in a suitable round flask at approximately 40 °C on the rotary evaporator (5.5). Any remaining traces of water are removed by adding ethanol (4.5) and evaporating again. The dry residue can, if necessary, be evaporated with a certain amount of n-hexane (4.4) to remove any traces of ethanol and is finally taken up in 3–5 ml n-hexane (4.4) (=carotene concentrate).

7.3. Separation of Carotene by Column Chromatography
Using a chromatography tube or an Allihn tube (5.6), the fritted plate is covered with a layer of sea sand approximately 0·5 cm thick (4.1) and a suspension of deactivated aluminium oxide (4.3.1) in n-hexane (4.4) is poured slowly in and the sedimenting particles distributed by lightly shaking the column. The aluminium oxide column, thus prepared, should be approximately 8–9 cm high and permanently covered with n-hexane. The carotene concentrate (7.2) is transferred completely by means of a dropping pipette with subsequent washing with small portions of n-hexane (4.4) on to the top of the column and the yellow-orange carotene which moves in front is subsequently eluted with n-hexane (4.4) in a suitable graduated flask, e.g. 100 ml. The process is terminated when the last fraction of eluate is shown spectrophotometrically to contain no more carotene colouring. The graduated flask is made up to the mark with n-hexane (4.4) (=test solution).

> *Note 10:* The column can take up to 200 µg total carotene without any deleterious effects on separating capacity. Accompanying carotenoids remain in the column—the xanthophylls at the top of the column, the lycopin in the middle.

7.4. Spectrophotometric Measurement

7.4.1. *Standard.* There is no requirement to use a standard since the calculations can be carried out with the experimentally determined and generally valid extinction of β-carotene in *n*-hexane.

$E_{1\,cm}^{1\%}$ = Extinction of a 1% β-carotene solution in *n*-hexane (in a 1 cm cuvette at the maximum near 450 nm) = 2590.

7.4.2. *Measurement.* Using *n*-hexane (4.4) as a blank solution, preferably in a 1 cm cuvette (5.8), the extinction of the test solution (7.3) is measured in the normal way on the spectrophotometer (5.7).

8. EVALUATION

8.1. Calculation

The total carotene content is calculated as β-carotene using the specific extinction of β-carotene in *n*-hexane (7.4.1) according to the equation:

$$\text{Total carotene (mg/100 ml (g) sample)} = \frac{\text{Ext} \times 100^a \times 1000 \times V_M \times V_P}{2590 \times 100^b \times E \times V_A}$$

where: Ext = extinction of the test solution (7.3) (scale divisions); 2590 = $E_{1\,cm}^{1\%}$ of β-carotene in *n*-hexane (scale divisions); E = weighed portion (6.2.1; 7.2) (ml(g)); V_M = volume of the test solution (7.3) (ml); V_P = volume of sample extract (7.2) (identical with volume of added chloroform) (ml); V_A = volume of the aliquot of the sample extract (7.2) (ml); 100^a = conversion to 100 ml (g) of sample; 100^b = conversion to 1 ml of test solution; 1000 = conversion to mg of carotene.

8.1.1. *Range of linearity.* The extinction is directly linear and proportional to the concentration in the range 100–5000 ng β-carotene/ml test solution.

8.1.2. *Detection limit.* A concentration of 100 ng/ml test solution can be quantitatively determined.

8.2. Reliability of the Method

8.2.1. *Recovery.* Recovery rates are 95–100%; it is not necessary to carry out determinations of the recovery rate or to correct these results.

8.2.2. *Repeatability.* This was not determined.

8.2.3. *Comparability.* This was not determined (no syndicate tests).

8.2.4. *Comparison with other carotene determination methods.* No details are available.

9. ANALYSIS REPORT

The result of the determination is to be given with a reference to this method (three procedures). Operations not mentioned in the described procedures must be indicated.

10. REFERENCES

Procedure A
10.1. U. Manz and J.-P. Vuilleumier, *Z. Lebensm. Unters.-Forsch.*, **163**, 21–4 (1977).
Bestimmung der pigmentierenden Carotinoide in Futtermitteln und Konzentraten für Eierproduktion und Geflügelmast.
10.2. DGF-Einheitsmethode F-II 1 (75) (1975).
Isolierung des Unverseifbaren.
10.3. DGF-Einheitsmethode F-II 2a (75) (1975).
Gewinnung und quantitative Bestimmung der Gesamtcarotine.
10.4. W. Müller-Mulot, *Fette-Seifen-Anstrichmittel*, **78** (1), 18–22 (1976).
Carotinanalysen in rohem Palmöl.
10.5. U. Manz, Roche Information Service, Animal Nutrition Department, F. Hoffmann–La Roche & Co. Ltd, Basle.
Methoden zur Bestimmung von β-Carotin in Rovimix-β-Carotin 10%, in Mischfuttern sowie von Carotin in Futtermitteln, Blutplasma und Milch (1981).
10.6. U. Manz, F. Hoffmann–La Roche & Co. AG, Basle (1981).
β-Carotin, Analysenmethoden, 14–17.
10.7. EWG 68-Verfahren Nr. 17 383/2/VI/68-D.
10.8. P. A. Jackson, *Food*, 12–18 (December 1982).
HPLC Analysis of Vitamins.

Procedure B
10.6.–10.8. See Procedure A.
10.9. R. Seibold and R. Bassler, Fachgruppe Futtermittel des VDLUFA (Verbandsmethoden) in Kraftfutter, **5**, 196 (1974).

10.10. Laboratory method, Control Department of Hoffmann-La Roche AG, Grenzach (FRG) (Ref. 10.7 modified) (since 1976).
10.11. A. Yang and L. Zhang, *Shipin Kexue (Beijing)* **35**, 18–25 (1982). Determination of Carotenes in food; Chinese research, similar in principle to Ref. 10.9.
10.12. Yen-Ping C. Hsieh and M. Karel, *J. Chromatogr.*, **259** (3), 515–18 (1983). Rapid extraction and determination of α- and β-carotenes in foods.

Procedure C

10.13. Laboratory method, Control Department of Hoffmann–La Roche AG, Grenzach (FRG) (1983).

4
Vitamin B₁ (Thiamine) in Foodstuffs: Thiochrome Method

1. PURPOSE AND SCOPE

The method describes a procedure for the quantitative determination of total vitamin B_1 (thiamine and its esters) in foodstuffs. It covers both naturally occurring vitamin B_1 as well as any vitamin B_1 that might have been added in the course of manufacture of the food. The method can be applied to fresh and stored foodstuffs intended for immediate consumption, such as cereals, flour, bread and confectionery, vegetables, fruit, potatoes, pasta, all meats, fish, milk and dairy products, jams and ready-to-eat meals and also to biological material.

Vitamin B_1 contents from as low as $50\,\mu g/100\,g$ are quantitatively measurable and contents as low as $20\,\mu g/100\,g$ can still be estimated semiquantitatively.

2. DEFINITION

Vitamin B_1 content is taken to mean the content of total vitamin B_1 as determined by the method described here and calculated as thiamine chloride hydrochloride. It is expressed in micrograms (µg) of thiamine chloride hydrochloride/100 g of sample.

3. BRIEF DESCRIPTION (PRINCIPLE OF THE METHOD)

After acidic disintegration and enzymatic digestion, in order to release the thiamine in the sample, the vitamin is largely freed of accompanying substances by the use of the weakly acidic cation-exchange resin Amberlite CG 50 I and then oxidised to the fluorescent derivative thiochrome. Naturally occurring mono-, di- and triphosphate esters of thiamine are co-determined as thiamine. The content is determined by means of a calibration curve plotted from matrix-free vitamin B_1 standard solutions which have undergone the same procedure as the sample. Furthermore, it is obligatory to carry out parallel recovery determinations with an internal vitamin B_1 standard added to the sample in amounts about equivalent to those originally present.

4. CHEMICALS

Remark: Unless otherwise specified, AR grade chemicals are to be used; water must be distilled from quartz glass vessels or be of equivalent purity.

4.1. Potassium hexacyanoferrate (III) (potassium ferricyanide), e.g. Merck.
4.2. Sodium hydroxide pellets (max. 0·0002% K), e.g. Merck.
4.3. Sodium acetate trihydrate, puriss., >99·5%, e.g. Fluka or Merck (quality: indifferent to $KMnO_4$).
4.4. Sodium chloride (common salt) cryst., purum, Ph.Eur., e.g. Fluka.
4.5. Quinine sulphate, $2H_2O$ purum DAB, fluorescent indicator, >99%, e.g. Fluka or Roth.
4.6 Thiamine chloride hydrochloride (thiamine HCl) (thiaminium dichloride) (vitamin B_1 hydrochloride) for biochemical purposes; drying loss approx. 3%, e.g. Merck.
4.7. Clara-Diastase (Clarase 300), Fluka, Art. No. 27540.
4.8. Ion-exchange resin Amberlite CG 50 I, 100–200 mesh (80–159 μm), analytical grade (highly purified), e.g. Serva, Art. No. 40511. Amberlite swollen in water can be kept in water for up to 6 months; swelling time at least 2 h.
4.9. Benzenesulphonyl chloride (benzenesulphonic acid chloride, benzenesulphone chloride) for synthesis, 99%, e.g. Merck.
4.10. Sulphuric acid, 95–97%, e.g. Merck.
4.11. Sulphuric acid, approx. 0·5 M (1·N) solution; from 27·5 ml H_2SO_4 (4.10) made up to 1000 ml with water. Where required, 0·4 M (0·8 N) and 0·2 M (0·4 N) solutions by dilution.
4.12. Sulphuric acid, approx. 0·1 M (0·2 N) solution; from 5·5 ml H_2SO_4 (4.10) made up to 1000 ml with water.
4.13. Hydrochloric acid, fuming, 37%, e.g. Merck.
4.14. Hydrochloric acid, approx. 1 M (1 N) solution; from 83 ml HCl (4.13) made up to 1000 ml with water.
4.15. Hydrochloric acid, approx. 0·15 M (0·15 N) solution; from 12·4 ml

HCl (4.13) made up to 1000 ml with water, or from 150 ml HCl (4.14) made up to 1000 ml with water.
4.16. Paraffin, low viscosity, DAB 8, USP.
4.17. Ethanol, absolute, e.g. Merck.
4.18. Isobutanol (isobutyl alcohol, 2-methyl-1-propanol), puriss., >99·5%, e.g. Fluka or Merck.
4.19. Petroleum ether, boiling range 50–75 °C.
4.20. Solutions
  4.20.1. Sodium hydroxide solution, 50% (w/v); dissolve 50 g NaOH (4.2) in water, while cooling, to make 100 ml.
  4.20.2. Potassium hexacyanoferrate (III) solution, 5% w/v; dissolve 5 g potassium hexacyanoferrate (III) (4.1) in water to make 100 ml; keeps at least 3 months in a cool place protected from light.

Note 1: The 5% concentration was chosen because reductive substances such as ascorbic acid, which may be present in foodstuffs, consume part of the potassium hexacyanoferrate (III). In all cases, once thiochrome formation has taken place (7.5.3) the alkaline phase must still be yellow; otherwise higher concentrations must be used. Since, however, an excess of potassium hexacyanoferrate (III) can destroy thiochrome, the optimum effective concentration has to be found. Bromine cyanide, which is equally effective as an oxidant, was too dangerous a substance to be taken into consideration.

  4.20.3. Sodium acetate solution, approx. 2·5 M; dissolve 340 g sodium acetate (4.3) in water to make 1000 ml.
  4.20.4. Clara-Diastase suspension, 10%; add water to 5 g Clara-Diastase (4.7) to make 50 ml and mix with a magnetic stirrer; prepare immediately before use; before using for the first time, test enzyme for natural thiamine content.
  4.20.5. Quinine sulphate fluorescent standard; dissolve 100 mg quinine sulphate (4.5) in H_2SO_4 (4.12) to make 1 litre (= stock solution; keeps for 3 months in refrigerator). Dilute 10·0 ml of stock solution with H_2SO_4 (4.12) to 100 ml; then dilute 10·0 ml of that with H_2SO_4 (4.12) to 100·0 ml (= *quinine sulphate fluorescent standard* with 1·0 µg quinine sulphate (4.5)/ml). Dilute solutions will keep for only one week; make fresh solution from stock solution as required and keep in refrigerator.

Note 2: The quinine sulphate fluorescent standard is used exclusively for adjusting the spectrofluorometer. It is also possible to use a fluorescent block in a similar way.

4.20.6. Vitamin B_1 standard solution; dry approx. 500 mg thiamine HCl (4.6) to constant weight at 60–70 °C in a vacuum (approx. 1–2 h).

Note 3: If the water content (drying loss) of thiamine chloride hydrochloride (4.6) is known, drying can be dispensed with and a conversion made to water-free thiamine chloride hydrochloride (4.6).

Dissolve 100 mg of the standard in H_2SO_4 (4.12) to make 1 litre ($=$ *vitamin B_1 stock solution*)—keeps for 2 months in a dark bottle (check contents). Dilute 10·0 ml of stock solution in H_2SO_4 (4.12) to 100·0 ml ($=$ *vitamin B_1 standard solution* with 10·0 μg thiamine chloride hydrochloride/100 ml). Prepare fresh solution before use.

5. APPARATUS AND AUXILIARY EQUIPMENT

5.1. Standard basic laboratory equipment.
5.2. Ground-glass Erlenmeyer flask, amber glass, standard joint 29/32, 100 ml, with stopper.
5.3. Stop watch.
5.4. Pipetting syringes up to 10 ml, e.g. Fortuna, Eppendorf.
5.5. Small autoclave or pressure cooker.
5.6. Water bath with thermostat.
5.7. pH meter or pH paper.
5.8. Polytron rod or similar mixing device.
5.9. Rotary evaporator.
5.10. Magnetic stirrer.
5.11. Chromatography tubes with capillary outlet, e.g. column 130–150 mm × 6 mm, reservoir 70–100 mm × 30–35 mm; capillary outlet 30–40 mm × 1 mm (see Fig. 10); also Bio-Rad Econo-Columns with 1·0 cm internal diameter and 6 cm packing height ($=4·7$ cm^3 column packing) and Luer three-way valve for eluant regulation.
5.12. Refrigerator.
5.13. Chest freezer.
5.14. Laboratory centrifuge with sealable centrifuge tubes (40 ml).
5.15. Pleated filter, e.g. Macherey & Nagel No. 616 1/4, diameter 18·5 cm.
5.16. Pleated filter, e.g. Macherey & Nagel No. 616 1/4, diameter 27 cm.
5.17. Cotton wool.

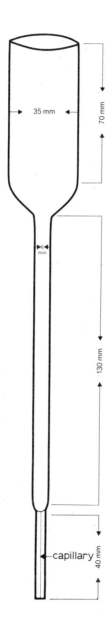

Fig. 10. Chromatography tube for chromatography of thiamine.

5.18. Spectrofluorimeter, commercial: excitation (Hg lamp) 365 nm, emission 435 nm; excitation (xenon lamp) 378 nm, emission 430 nm.

6. SAMPLE

6.1. Sampling

The proportional composition of the sample taken must be representative of the material (foodstuff) to be analysed. The initial quantity of sample must always be sufficiently large.

6.2. Sampling Method and Preparation of Samples

If it is necessary to store foodstuffs before analysis can be performed, perishable, freshly sampled foodstuffs for field tests, etc., have to be preserved as follows. The sample material (which may be deep frozen for short-term storage, but if longer storage is required the sample should then be carefully defrosted at $+4\,°C$) is taken up in 0·4 M, 0·2 M or 0·1 M sulphuric acid (4.11 and 4.12) and homogenised with the Polytron rod (5.8) between 20 s and 2 min depending on the type of material.

The quantity ratio of sample material to sulphuric acid must be such that homogenisation yields a freely flowing medium with a final acid concentration of approx. 0·1 M. For example, for 90–100 g samples of soups, juices, milk and other materials consisting largely of water, 0·4 M sulphuric acid is to be used; 0·2 M for 60–70 g samples of meat, vegetables, fruit, etc.; and 0·1 M sulphuric acid for 20–30 g samples of bread, flour, baked confectionery and other highly swelling products containing little water. For dried products which swell considerably, e.g. crispbread, the dilution must be even greater. The sulphuric acid solutions are then deep frozen at $-20\,°C$ in which condition they will keep for up to three months.

Samples which do not require preservation treatment, e.g. dry, imperishable foodstuffs, are comminuted and homogenised and then immediately placed in 0·1 M sulphuric acid (4.12) for disintegration and digestion. Highly fatty (dry) samples are preferably defatted with petroleum ether (4.19) before being taken up in sulphuric acid. This operation can also be done with the weighed portion itself.

6.2.1. Weighed portion.
The weighed portion from the sample material prepared in accordance with method 6.2 must be so chosen that an

aliquot of the sample extract (7.2.2) of not more than 20 ml will contain approx. 3–5 µg of thiamine chloride hydrochloride; the quantity of sample material in the weighed portion, i.e. without acid, is of the order of 10–20 g. As a rule, four weighed portions must be prepared from each sample preparation: two portions for determining the amount of vitamin B_1 present originally and two portions for recovery determinations to be carried out in parallel (genuine repeat determination in each case).

7. PROCEDURE

7.1. Remark
The analysis must be carried out quickly at every stage under subdued light (possible interruption points are indicated); repeat determinations must be made on the same day.

7.2. Disintegration and Digestion

7.2.1. *Disintegration with sulphuric acid.* Sample material (6.2) which has not undergone preservation treatment but which may have been defatted is transferred in a weighed portion, standardised according to method 6.2.1, into a 100 ml graduated flask, and approx. 40 ml of 0·1 M sulphuric acid (4.12) is added; the mixture has to be highly fluid.

Preserved, deep-frozen sample material (6.2) is carefully brought up to room temperature. After it has been homogenised again, the weighed portion (standardised according to method 6.2.1) is transferred into a 100-ml graduated flask. Enough 0·1 M sulphuric acid (4.12) is added to bring the total quantity of acid to approx. 40 g. A check must always be kept on the pH value of the homogenised sample diluted with acid in the graduated flask (e.g. sample material containing carbonate or alkali will need to be neutralised carefully and the pH adjusted); the pH value should be ≤ 2. The graduated flask is lightly stoppered and the contents are treated in the small autoclave (5.5) for 15 min at 1 atm excess pressure and 120°C. Warning: beware of spray! If necessary, add about 3 ml of paraffin (4.16) before autoclaving.

For the recovery determinations, an amount of thiamine chloride hydrochloride equivalent to the original value to be expected (or already found) is added to make the parallel mixture—appropriate quantities being taken from the vitamin B_1 stock solution or standard solution (4.20.6). The volume added should in principle be kept small.

Note 4: The analytic procedure is adapted to thiamine chloride hydrochloride quantities of approx. 3–12 μg in view of the column loading. Recovery determinations with added vitamin B_1 are used for determining the correction factor for matrix-induced vitamin B_1 losses.

7.2.2. *Enzymatic digestion* (*obtaining the extract*). When the autoclave has cooled down to 35–40 °C, the samples are taken out and sufficient sodium acetate solution (4.20.3) added to them, while shaking, to bring the pH of the suspension to a value between pH 4·0 and 4·5 (approx. 5 ml). Five millilitres of Clara-Diastase suspension (4.20.4) is added and the samples are incubated for 20 min or longer in a water bath at 45 °C. The sample is brought to room temperature, made up with water to 100·0 ml (any paraffin present must lie above the mark), mixed and filtered through a pleated filter (5.15) (discard the first runnings). This removes from the extract not only the coarse parts but also any fats not previously extracted (6.2) (=raw sample extract).

Note 5: For determination of unesterified thiamine (naturally occurring and possibly added), enzymatic digestion is omitted. The cooled sample (7.2.1) is brought to pH 4·0–4·5 with sodium acetate solution (4.20.3), made up to the mark with water, mixed and filtered.

7.3. Purification of the Extract

7.3.1. *Removal of starch and proteins* (preliminary purification, if required). If samples are rich in starch and protein, the filtered raw sample extract (7.2.2) must be pre-purified by alcohol precipitation before the ion-exchange column is used. An aliquot (e.g. 50 ml) of the sample extract is taken up in 10 (or even 15 or 20) times as much ethanol (4.17), thus precipitating out the starch and proteins. After a short wait (to allow the precipitate to settle) these substances are separated over a G4 glass filter or a pleated filter (5.16) while washing out with ethanol (4.17). All the ethanol is driven off the filtrate by heating at max. 40 °C under vacuum in the rotary evaporator.

Note 6: Removing the ethanol always concentrates the pre-purified extract, since water is removed at the same time (syrupy residue). This can be exploited as a means of concentrating vitamin B_1.

The syrup which remains is transferred into a 50 ml (see above) graduated flask by washing out with water several times and made up to the mark, the pH value being again adjusted to pH 4·0–4·5 (=sample extract pre-purified and freed of fats, starch and protein).

7.3.2. *Purification over cation-exchange resin.* Amberlite CG 50 I (4.8), which has previously been soaked in water, is freed of fine particles by repeatedly pouring off the surface water. A column (5.11) is packed to a height of 7 cm with the ion-exchange resin thus prepared. The drip rate of the packed column should be 15–20 drops/min; it can be regulated by placing wadding in the junction between the outlet capillary and the column and by varying the packing density of the Amberlite. The ion exchange resin is then activated with 50 ml of 1 M hydrochloric acid (4.14) and neutralised by washing with water (pH check with universal indicator paper). A transfer pipette is now used to place enough (max. 20 ml) of the raw sample extract (7.2.2) or of the pre-purified sample extract (7.3.1) on the column for about 3–5 μg of thiamine HCl to enter the Amberlite; 25 ml of water is used to wash out in 1×5 ml and 2×10 ml portions.

Remark: The preparations for recovery determination contain roughly twice the amount of vitamin B_1 in the extracts, and for this reason only half the volume of the corresponding sample extract aliquot is applied to the column.

The thiamine base bound to the ion exchanger is now eluted with 25 ml of 0.15 M hydrochloric acid (4.15) applied in portions of 1×10 ml and 1×15 ml (allow column to run dry), the eluate being collected in a 50-ml graduated flask and brought up to the mark with 0.15 M hydrochloric acid (4.15) (=purified extract containing approx. 3–5 μg of thiamine HCl=sample solution). Four sample solutions are obtained for each sample material (see section 6.2.1).

Note 7: The ion-exchange resin is used only once and then discarded.

7.4. *Standards for Calibration Curve*

To plot a calibration curve, the analysis method described in sections 7.2.1, 7.2.2 and 7.3.2 with 20 ml of the solution described in section 7.2.2 placed on the column must be carried out in duplicate, matrix-free, with 1.0, 1.5, 2.0, 2.5 and 3.0 ml of vitamin B_1 standard solution (4.20.6), equivalent to 10, 15, 20, 25 and 30 μg of thiamine HCl (=calibration standards with 2.0–6.0 μg thiamine HCl/50 ml).

7.5. *Oxidation of Thiamine to Thiochrome*

7.5.1 *Remark.* The oxidation of thiamine to thiochrome and the extraction of the latter must be carried out with strict adherence to the conditions specified in section 7.5.3 (quantities, volumes, times, temperature and stirring rate). Rigorous attention to detail is clearly necessary in this procedure. The procedure is carried out on:

(a) 10 calibration standards (7.4) with 2·0–6·0 µg thiamine HCl/50 ml;
(b) two sample solutions (7.3.2) with approx. 3–5 µg thiamine, HCl/50 ml;
(c) two sample solutions for recovery determination (7.3.2) with approx. 3–5 µg thiamine HCl/50 ml.

The reactions are performed uniformly in 100-ml ground-glass Erlenmeyer flasks made of amber glass with ground stoppers (5.2), while stirring with a magnetic stirrer (Teflon rods).

Note 8: The quantity of thiamine HCl is limited to 3–5 µg/50 ml of sample solution in order to keep within the upper and lower limits of the calibration curve when measuring the solutions, and to avoid having to change to another measurement sensitivity and thus to be more sure that the measurements are accurate.

7.5.2. *Preparation for oxidation.* The 10 calibration standards (7.4) and the four sample solutions (7.3.2), each of 50 ml, are halved:

(a) for unblocked oxidation (main value), remove 25·0 ml of the solution with a transfer pipette and place it in a 100-ml ground-glass Erlenmeyer flask.
(b) for blocked oxidation (blank), transfer the remaining 25 ml of the solution into a 100-ml ground-glass Erlenmeyer flask by pouring out without rinsing.

The 28 solutions of 25·0 ml each which have been prepared for oxidation must all be uniformly subjected to the further treatment described in section 7.5.3 before an hour has elapsed. They contain 1·0–3·0 µg thiamine HCl (halved calibration standards) or approx. 1·5 – 2·5 µg thiamine HCl (halved sample solutions) respectively.

7.5.3. *Oxidation and thiochrome extraction.* The solutions prepared according to method 7.5.2 are subjected to oxidation treatment according to Table 2; the thiochrome formed is extracted at the same time. A stop watch (5.3) is used.

When oxidation is complete, the isobutanol extract must be further processed immediately if possible, but it will keep for 1 h over the alkaline phase, including centrifuging. The reaction mixture, principally its isobutanol phase, is transferred into centrifuge tubes (5.14) and centrifuged at 4000 rpm for 2–3 min. Ten millilitres of the supernatant isobutanol phase is pipetted (transfer pipette) into a ground-glass test tube, 0·5 ml of absolute ethanol (4.17) is added and mixed in after the test tube has been sealed. The solution may be left standing for 1 h until measurement begins if protected from light. When the calibration curve is plotted at the same time, 28 solutions of 10·5 ml each are obtained, 14 for main values and 14 for blanks (=calibration and sample test solutions).

TABLE 2

Sequence of additions	Unblocked oxidation (main value)	Blocked oxidation (blank)
Sample solutions and calibration standards (7.5.2)	25·0 ml	25·0 ml (remainder of solution)
Isobutanol (4.18)[a]	15·0 ml	15·0 ml
Sodium hydroxide solution (4.20.1)[b]	3·0 ml	3·0 ml
Waiting time (from beginning of NaOH addition)	10 s	10 s
Benzenesulphonyl chloride (4.9)	None	0·3 ml
Waiting time	None	50 s
Potassium hexacyanoferrate (III) solution (4.20.2)	0·65 ml	0·65 ml
Waiting time	60 s	60 s
Sodium chloride (4.4)[c]	6·0 g	6·0 g
Increased stirring rate[d]	60 s	60 s

[a] Stirring rate: set magnetic stirrer (5.10) so that phases are thoroughly mixed.
[b] Fast input with pipetting syringe (5.4).

> *Note 9:* Adding the sodium hydroxide solution *before* oxidation results in 6–8% higher thiochrome yields than using ready-made oxidation solution (sodium hydroxide solution + potassium hexacyanoferrate (III) solution). If the latter is used, it must be added quickly; if the addition takes 20 s, the thiochrome yield will be reduced by about 30%.

[c] Input through powder funnel with large internal diameter.

> *Note 10:* Salting-out effect! Do not add until oxidation is complete. If sodium chloride is present during oxidation, the thiochrome yield will be reduced by 15–19%. See also Ref. 10.8.

[d] During this time seal the Erlenmeyer flask.

7.6. Fluorimetric Measurement

7.6.1. *Adjustment of the spectrofluorimeter.* The measuring instrument (5.18) can be adjusted with the quinine sulphate fluorescent standard (4.20.5) or alternatively with a suitable fluorescent block. First, adjust setting to 0 scale divisions against air, then to 70 divisions (for example) against the standard. The procedure for digital indication is the same. The scale range should be so adjusted that 1·5 µg of thiamine HCl in 25·0 ml of the calibration standard (7.4) gives a fluorescence reading of about 30 scale divisions. Wavelength settings for maximal sensitivity for a Hg lamp are 365 nm (excitation) and 435 nm (emission); the values for a xenon lamp are 378 nm (excitation) and 430 nm (emission) (see section 5.18). Only cuvettes with the same self-absorption (QI type) may be used.

7.6.2. *Plotting of the calibration curve (with calibration test solutions).* The calibration curve is plotted from the measured values of the calibration test solutions (main reading minus blank reading, in each case as the mean from duplicate determinations) obtained for 1·0, 1·5, 2·0, 2·5 and 3·0 µg of thiamine HCl (7.5.3). In the given concentration range it is linear and proportional to the quantity. The calibration curve is used for evaluating the samples to be analysed.

> *Note 11:* The fact that the calibration curve has so far always proved to be linear and proportional within the specified calibration range is sufficient justification for dispensing with the calibration curve if the thiamine HCl concentration in 25·0 ml of the sample solution (7.3.2) is very close to that of a halved calibration standard (7.4) with a concentration of about 1·5 µg/25·0 ml (working with one calibration point). For series analyses of the same sample material, for example, this simplification is permitted.

7.6.3. *Measurement of sample test solutions.* The measured values (main reading minus blank reading) of the sample test solutions (7.5.3) are likewise to be given as the mean values from true duplicate determinations (see 6.2.1). The actual measurement should be performed several times (sufficient quantities of sample test solutions (10·5 ml) are available).

8. EVALUATION

8.1. The vitamin B_1 content of the sample material is calculated according to the following formula:

$$\mu g \text{ thiamine HCl}/100 \text{ g} = \frac{A \times \text{Dil} \times 100^a}{E_p} \times \frac{100^b}{\text{Rec}}$$

where: A = thiamine HCl in 25·0 ml of sample solution (7.5.2), read from calibration curve or calculated from a single calibration point measurement according to the formula:

$$A = \frac{(\text{measured value for sample}) \times (y \, \mu g \text{ thiamine HCl})}{(\text{measured value for standard for } y \, \mu g \text{ thiamine HCl})} (\mu g)$$

(see Note 11); Dil = dilution factor, to be determined individually in each case and to be doubled for the corresponding recovery mixtures; E_p = quantity of original sample material (6.2.1) in the weighed portion (g); 100^a = conversion to 100 g of original sample material; Rec = recovery rate (%) (mean of two determinations), calculated from:

$$\text{Rec} = \frac{(\text{value found after addition} - \text{value found without addition})}{\text{theoretical value of the addition}} \times 100$$

100^b = conversion to corrected content (%).

8.1.1. *Range of linearity.* In the range 1–3 µg of thiamine HCl per 25·0 ml calibration standard (7.4) the emission is linear and directly proportional to concentration (results of many checks; see also Note 11).

8.1.2. *Detection limit.* There is a quantitatively measurable indication for as little as 0·033 µg of thiamine HCl/ml of isobutanol (approx. 10 scale divisions). Qualitative detection is at approx. 15 µg/100 g.

8.2. *Reliability of the Method*

8.2.1. *Recovery.* About two-thirds of all foodstuffs so far analysed by this method showed recovery rates of 95–100% in the vitamin B_1 determination. In the case of products containing sugar and cocoa, values of 60–80% have occasionally been found, and likewise in certain types of bread and dairy products. Values between 91 and 98% have been obtained for plain flour, and in the case of vegetables the recovery rates have been up to 100%.

It is therefore essential to carry out recovery determinations in parallel. The vitamin B_1 losses are clearly due to the matrix; they can occur during disintegration (chemical, enzymatic), purification with Amberlite, oxidation (inhibitory effect) and during measurement (fluorescence quenching).

8.2.2. *Repeatability.* The relative standard deviation in series of repeat determinations carried out on various foodstuffs with vitamin B_1 contents of 20–1000 µg/100 g of foodstuff was around 2·87%. The following values were obtained from vitamin B_1 analyses of meat products:

boiling sausage 270 µg/100 g; for $n=5$: $s_{rel} = 3·81\%$;
freeze-dried product 3000 µg/100 g; for $n=4$: $s_{rel} = 1·64\%$;
freeze-dried product 5000 µg/100 g; for $n=4$: $s_{rel} = 0·98\%$.

In yeast, rusk and half plain flour (thiamine contents between 0·3 and 17 µg/100 g) relative standard deviations from 4·2 to 6·5% were found.

8.2.3. *Comparability.* The method described has not yet been tested in a syndicate test.

8.2.4. *Comparison with other vitamin B_1 determination methods.* In two syndicate tests of meat products carried out by the Bundesanstalt für Fleischforschung in Kulmbach (seven participants in 1980, nine in 1982) there was very close agreement between the results obtained by the method described and those obtained by the Bundesanstalt method, which also uses the Amberlite column. Users of Decalso (in place of Amberlite) have found 10–12% less vitamin B_1.

9. ANALYSIS REPORT

The result of the determination is to be given with a reference to this method. Operations not mentioned in the described method have to be indicated.

10. REFERENCES

The following literature sources and communications have been taken as the basis for the method described or given substantial background information:

10.1. B. C. P. Jansen, *Rec. Trav. Chim. Pays-Bas,* **55,** 1046–52 (1936).

10.2. R. Rettenmaier, J.-P. Vuilleumier and W. Müller-Mulot, Z. Lebensm. Unters. Forsch., **168**, 120–4 (1979).
10.3. Bundesanstalt für Fleischforschung, Institut für Chemie und Physik, D-8650 Kulmbach, Syndicate tests 1981 and 1982.
10.4. ICC-Standard No. 117 in Arbeitsgemeinschaft Getreideforschung Standardmethoden für Getreide, Mehl und Brot, 6. Auflage, Verlag Moritz Schäfer, Detmold (1978).
10.5. W. Müller-Mulot, GDCh-Fortbildungskurs Detmold, 11.3.1982 (lecture).
10.6. J. Schütt, *Lebensmittelindustrie*, **26** (1), 21–4 (1979).
Zur Vitamin B-Bestimmung in pflanzlichen und diätetischen Lebensmitteln.
10.7. B. Gassmann, D. Lexow. and V. Erhardt, *Nahrung*, **16** (3), 245–57 (1972).
Problematik und Grenzen der Standardisierung von Methoden zur Vitamin-Bestimmung in Lebensmitteln.
10.8. T. Fukuda and K. Kobayashi, *Bitamin*, **55** (9–10), 445–51 (1981).
Effect of NaCl on the thermal decomposition of thiamine in the neutral range.
10.9. A.-G. M. Solimann, *J. Assoc. Off. Anal. Chem.*, **64** (3), 616–22 (1981).
Comparison of manual and benzenesulfonyl chloride-semiautomated thiochrome methods for determination of thiamine in foods.
10.10. B. Poth and H. Dahlke, *Bäcker und Konditor*, **34** (10), 294–6 (1980).

On the separation of thiamine and its three phosphate esters:

10.11. K. Ishii, K. Sarai, H. Sanemori and T. Kawasaki, *J. Nutr. Sci Vitamimol.* **25**, 517–23 (1979).
10.12. M. Kimura, T. Fujita, S. Nishida and Y. Itokawa, *J. Chromatog.*, **188**, 417–19 (1980).
10.13. M. Kimura, B. Panijpan and Y. Itokawa, *J. Chromatog.*, **245**, 141–3 (1982).

5
Vitamin C (Ascorbic and Dehydroascorbic Acids) in Foodstuffs: HPLC Method

1. PURPOSE AND SCOPE

A method is described for the quantitative determination of vitamin C in foodstuffs. This method detects the total vitamin C (ascorbic acid and dehydroascorbic acid) of natural origin; ascorbic acid added during the processing of the foodstuff may also be detected. The method does not make a distinction between ascorbic acid and erythorbic acid. The method is applicable to both fresh and stored foodstuffs intended for direct consumption such as vegetables, fruit, meat, milk and milk preparations. Vitamin C contents of over 0·5 mg/100 g sample can be determined quantitatively.

2. DEFINITION

Vitamin C content is understood to be the content of ascorbic acid plus dehydroascorbic acid, calculated as ascorbic acid, determined by the method described here. It is given in mg ascorbic acid/100 g sample.

3. BRIEF DESCRIPTION (PRINCIPLE OF THE METHOD)

After reduction, the material under investigation is homogenised in cold 4% metaphosphoric acid in a kitchen mixer (Waring blender). In the case of samples with a high water content or of liquid samples a more concentrated metaphosphoric acid is used so that the homogenate still contains 4% of this acid. Dehydroascorbic acid is converted into ascorbic acid by means of 2,3-dimercapto-1-propanol (BAL). The sample extract, obtained after centrifuging and filtering the homogenate, is chromatographed without preliminary clean-up on an anion exchange column by means of HPLC after adjusting to a suitable concentration. Evaluation is carried out by comparing the peak heights against an ascorbic acid standard.

4. CHEMICALS

> *Remark:* Unless otherwise indicated AR grade chemicals are to be used; double distilled water or water of corresponding purity contained in quartz glass vessels must be used.

4.1. Metaphosphoric acid, rods, $60\% \pm 2\%$ (HPO_3), $36\% \pm 1\%$ ($NaPO_3$), e.g. Carlo Erba; approx. 65% (HPO_3), approx. 35% ($NaPO_3$), e.g. Fluka No. 79615 or Prolabo No. 20632.

> *Note 1:* Metaphosphoric acid with a higher content of $NaPO_3$ (slightly soluble, vitreous lump quality) should not be used in this case.

4.2. Sodium chloride (cooking salt) crystal, pure, Ph. Eur., e.g. Fluka.
4.3. Sodium hydroxide, pellets (maximum 0·0002% K), e.g. Merck.
4.4. β-Alanine (3-aminopropionic acid), highest purity, over 99%, e.g. Fluka.
4.5. L (+)-Ascorbic acid (vitamin C), AR and for biochemical purposes, e.g. Merck.
4.6. 2,3-Dimercapto-1-propanol (BAL) for complexometry, 98·5%, e.g. Merck.
4.7. Hydrochloric acid, 32%, e.g. Merck.
4.8. Nitrogen, oxygen-free, 99·9% (v/v).
4.9. Solutions.
    4.9.1. Metaphosphoric acid solution, 4%; dissolve 40 g of metaphosphoric acid (4.1) in water, make up to 1000 ml and mix with 1 ml BAL (4.6); can be stored for seven to ten days when frozen (see Remark 7.1).
    4.9.2. Metaphosphoric acid solution, 8%; dissolve 80 g of metaphosphoric acid (4.1) in water, make up to 1000 ml and mix with 1 ml BAL (4.6).
    4.9.3. Hydrochloric acid, approx. 10%; 1 volume of HCl (4.7) +2 volumes of water.
    4.9.4. Hydrochloric acid, approx. 1%; 1 volume of HCl (4.9.3) +9 volumes of water.
    4.9.5. Caustic soda, approx. 12%; 12 g NaOH (4.3) dissolved in water, made up to 100 ml.

68 METHODS FOR THE DETERMINATION OF VITAMINS IN FOOD

4.9.6. Ascorbic acid standard solution; dissolve 10·0 mg ascorbic acid (4.5) in metaphosphoric acid (4.9.1), make up to 100 ml (=stock solution), ascorbic acid content 100 µg/ml; make up new daily: 1 volume part stock solution diluted with 9 volume parts water (=ascorbic acid standard solution with 10 µg ascorbic acid/ml).

4.9.7. Mobile phases for HPLC: aqueous solution containing, per litre, 0·1 mol (=8·91 g) β-alanine (4.4), 0·1 mol (=5·85 g) sodium chloride (4.2) and 0·1% (=1·0 ml) BAL (4.6), adjusted with hydrochloric acid, approx. 10% (4.9.3), to a pH value of 4·5.

5. APPARATUS AND AUXILIARY EQUIPMENT

5.1. Usual basic laboratory equipment.
5.2. Thermostatically controlled water bath.
5.3. Thermostat (for column heating).
5.4. Laboratory pump (for filling columns), e.g. Technikon.
5.5. Equipment for the division and reduction of the sample material: knives, shears.
5.6. Kitchen mixer for homogenising samples (Waring blender).
5.7. Micropipettes, 10–500 µl.
5.8. Laboratory centrifuge.
5.9. Pleated filters (for homogenate).
5.10. HPLC apparatus consisting of:
Pump, e.g. Milton Roy Minipump.
Liquid chromatograph, e.g. Carlo Erba model 9100.
UV detector 254 nm.
Flow-through cell 10 × 1 mm (8 µl).
Recorder.
Column: glass column, 35 × 0·4 cm, with jacket, thermostatically controlled at 55 ± 0·5 °C; filled with anion exchanger Aminex-A 14, 20 ± 3 µm, Cl⁻ form, Bio-Rad Laboratories (Richmond, California, USA); filled to 30 cm after pre-treatment, corresponding to approx. 4–5 g of Aminex-A 14.

5.10.1. *Pretreatment of anion exchanger and filling of the column.* Aminex-A 14 (Cl⁻) is suspended in 10% hydrochloric acid (4.9.3) for 10 min at 80 °C on the water bath. After pouring into a Büchner funnel the column material is washed in succession with water, 12% caustic soda (4.9.5), water, 10% hydrochloric acid (4.9.3), water and finally 1% hydrochloric acid (4.9.4). The exchanger is again suspended in 1% hydrochloric acid (4.9.4), degassed under vacuum at 55 °C, poured into the glass columns and packed down to 30 cm with 1% hydrochloric acid (4.9.4) using the pumps (5.4) at a flow rate of 100 ml/h. The column temperature is maintained at 55 °C. The final conditioning of the exchanger is effected by pumping through 1% hydrochloric acid (4.9.4) for 3–4 h at a flow rate of 100 ml/h. In each treatment stage (hydrochloric acid, water, caustic soda, etc.) a volume of the washing solution equal to 10–20 times the volume of the exchanger is to be used (reagent excess).

Note 2: The column package should be regenerated after 50 to (maximally) 100 analyses. The procedure is identical with the pretreatment described for the anion exchanger. There has been no evidence of fusion, lumping together or clogging of the column filling material (chromatography at 55 °C). The use of pre-columns is not necessary. Regeneration is possible on packed columns.

6. SAMPLE

6.1. *Sampling*

The sample to be taken must be representative of the material to be investigated (foodstuffs) with regard to proportions and composition. In any case a sufficiently large initial quantity should be available. Frozen samples must be analysed after 1 week at the latest.

6.2. *Sampling Technique and Preparation of the Sample*

Structured and inhomogeneous material must be comminuted (5.5) to small pieces before the sample is taken for disintegration, homogenisation and extraction (5.6; 7.2). The equipment must be exclusively of glass, ceramic or enamel, with the exception of knives and shears.

6.2.1. *Weighed portion.* The weighed portion should be so chosen that when the homogenate (sample extract) has been filtered and made up to the mark, the resulting sample test solution (7.3) contains 5–30 µg ascorbic acid/ml.

7. PROCEDURE

7.1. *Remark*

When in solution, vitamin C is sensitive to oxygen, oxidising agents and catalytically active traces of metal. The addition of 2,3-dimercapto-1-propanol (BAL) has the effect (a) of converting traces of metal into inactive mercaptides, (b) of reducing any dehydroascorbic acid to ascorbic acid, (c) of protecting ascorbic acid from oxidation during the preparation of the sample and the chromatographic analysis. The material to be analysed should be prevented from coming into contact with metals (steel, copper, zinc equipment).

> *Note 3:* Dithiothreitol (1,4-dimercapto-2, 3-butanediol) produced by the companies Fluka and Serva has the same effect as BAL; see Refs 10.2 and 10.16.

7.2. Disintegration and Extraction

After preparation (6.2) and weighing (6.2.1), the weighed portion of food is homogenised in a food mixer (5.6) in a suitable volume of cold 4% metaphosphoric acid (4.9.1). In the case of products with a high water content or liquid products 8% acid (4.9.2) is used so that the final concentration of metaphosphoric acid also amounts to 4%. The homogenate is centrifuged conventionally at 5000–6000 rpm (5.8), and the upper phase filtered (5.9) (=sample extract). This extract should be further processed immediately; any freezing, even for a short time, may result in ascorbic acid losses.

> *Note 4:* Deep freezing should be limited to $-40\,°C$ and a maximum of 1 week. Sample extracts stored for 8 weeks exhibited slight vitamin C losses.

7.3. Sample Test Solution

The sample extract (7.2) is made up to the mark in a suitable measuring flask with 4% metaphosphoric acid (4.9.1) so that 5–30 mg ascorbic acid are contained in 1 litre, corresponding to 5–30 µg/ml (=sample test solution).

7.4. Standard Test Solution

Using, progressively, 50–100 µl of ascorbic acid standard solution (4.9.6), corresponding to 0·5–1·0 µg ascorbic acid, matrix-free chromatographic calibration is carried out. The standard peak heights obtained are taken to correspond to the theoretical amounts of ascorbic acid in the standard solution (4.9.6).

> *Note 5:* The measurement range of 0·5–1·0 µg ascorbic acid is determined by the high sensitivity of the detector. The quantity of sample test solution (7.3) which reaches the column should be adapted to the standard measurement range (see section 7.5).

7.5. HPLC

7.5.1. The complete separation and the measurement of the ascorbic acid is effected by means of the anion exchange HPLC. The total ascorbic acid is determined in this process. Following the chromatographic treatment of four sample test solutions (7.3), a standard test solution (7.4) with a suitable concentration should be injected.

Note 6: Ascorbic acid and isoascorbic acid (erythorbic acid) are not separated (the latter is always added). Any isoascorbic acid which may be present can be determined separately according to Refs 10.13–10.19.

7.5.2. Chromatographic conditions:

Stationary phase: Aminex-A 14, pretreated (5.10.1).
Mobile phase: See section 4.9.7.
Volumes and quantities: *For sample test solution (7.3):*
From 10 µl and up, to a maximum of 500 µl, depending on vitamin C content, with ascorbic acid quantities between 0·5 and 1·0 µg.

Note 7: Examples of the volumes of the sample test solution: 10 µl for green peppers, 500 µl for milk.

For standard test solution (7.4):
A microlitre quantity with a suitable concentration selected from the calibration scale (between 50 and 100 µl, corresponding to 0·5–1·0 µg ascorbic acid). Every 5th chromatographic analysis should be done with the standard test solution.

Detection: By means of the retention time for ascorbic acid and/or co-eluted added reference ascorbic acid, at $\lambda = 254$ nm.
Peak evaluation: By peak height.
Flow rate: 1 ml/min.
Pressure: 250 psi.
Sensitivity: 0·1 AUFS.

A standard chromatogram and examples of chromatograms for tomatoes and parsley are given in Fig. 11.

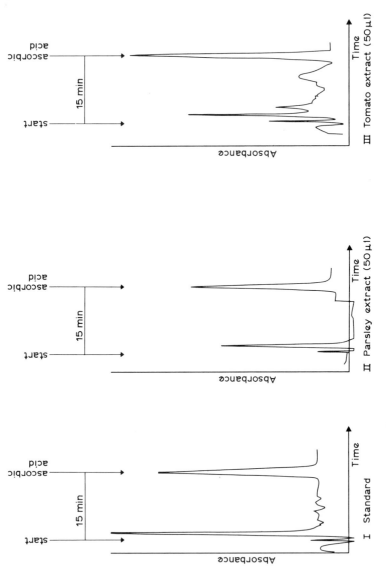

Fig. 11. HPLC chromatograms of ascorbic acid containing extracts.

8. EVALUATION

8.1. Calculation

The total vitamin C content of the original material to be investigated is calculated according to the formula:

$$\text{mg ascorbic acid}/100\,\text{g} = \frac{H_P C_S V \times 100}{H_S q a \times 1000}$$

where: H_S = peak height for ascorbic acid in standard test solution (7.4) (mm); H_P = peak height for ascorbic acid in sample test solution (7.3) (mm); C_S = ascorbic acid concentration in standard test solution (7.4) (µg); V = total volume of the sample test solution (7.3) containing the total sample extract (7.2) (ml); q = weighed portion (6.2.1) (g); a = volume of injected sample test solution (7.3) (ml); 100 = conversion to 100 g sample; 1000 = conversion to mg ascorbic acid.

8.1.1. *Range of linearity.* In the measurement range 0·5–5·0 µg ascorbic acid the peak heights are linearly proportional to the concentrations.

8.1.2. *Detection limit.* This amounts to 5 µg/ml for the standard test solution.

8.2. Reliability of the Method

8.2.1. *Recovery.* The recovery rate of ascorbic acid for Aminex-A 14 columns amounted to $99 \pm 2\%$. The recovery rates for vitamin C in the sample materials amounted to 90–92%. It is common practice not to correct the analysis result for these losses.

8.2.2. *Repeatability.* The relative standard deviation for this method is not indicated. For data on precision see Ref. 10.1.

8.2.3. *Comparability.* This was not carried out for this method (no syndicate tests).

8.2.4. *Comparison with other methods for determination of vitamin C.* The method described above was compared with three independent methods from the literature on the same samples of foodstuffs. The agreement was very good; see review of literature in Ref. 10.1.

9. ANALYSIS REPORT

The result of the determination is to be given with a reference to this method. Operations not mentioned in the described method have to be indicated.

10. REFERENCES

A number of papers have been published on the quantitative determination of vitamin C in foodstuffs using HPLC, mainly since 1980. They describe determinations made using reverse-phase ion pair chromatography and anion exchange chromatography. The method given above is based on the work quoted in Refs 10.1 and 10.11 and on private communications from Professor Fidanza in Perugia.

10.1. A. Floridi, R. Coli, A. Alberti Fidanza, C. F. Bourgeois and R. A. Wiggins, *Int. J. Vit. Nutr. Res.*, **52**, 194–7 (1982).
10.2. G. L. Lookhart, S. B. Hall and K. F. Finney, *Cereal. Chem.*, **59** (1), 69–71 (1982).
10.3. R. W. Keating and P. R. Haddad, *J. Chromatog.* **245**, 249–55 (1982).
10.4. A. E. Watada, *Hort. Sci.*, **17** (3), 334–5 (1982).
10.5. D. B. Dennison, T. G. Brawley and G. L. K. Hunter, *J. Agric. Food Chem.*, **29** (5), 927–9 (1981).
10.6. J. Augustin, C. Beck and G. J. Marousek, *J. Food Sci.*, **46**, 312–13, 316 (1981).
10.7. J. W. Finley and E. Duang, *J. Chromatog.* **207**, 449–53 (1981).
10.8. N. Miki, *Nippon Shokuhin Kogyo Gakkaishi*, **28** (5), 264–68 (1981).
10.9. H. Rückemann, *Z. Lebensm. Unters. Forsch.* **171**, 357–9 (1980).
10.10. L. Manso, *EFCE Publ. Ser.* **1980**, 12, 137–44. (Proc. Congr. Nac. Quim., 3rd, 1980, Vol. 2.)
10.11. A. Floridi, C. A. Palmerini and C. Fini, *J. Chromatog.* **138**, 203–12 (1977).
10.12. S. P. Sood, L. E. Sartori, D. P. Wittmer and W. G. Haney, *Anal. Chem.*; **48** (6), 796–8 (1976).
10.13. G. Parolari, *Industria Conserve*, **57**, 19–22 (1982).

The separation of ascorbic and isoascorbic acids by means of HPLC is described in:

10.14. J. Hofman, R. van Dijk, G. de Vries and G. van de Haar, *De Waren-Chemicus*, **11**, 84–90 (1981).
10.15. J. M. Coustard and G. Sudraud, *J. Chromatog.* **219** (2), 338–42 (1981).
10.16. L. W. Doner and K. B. Hicks, *Analyt. Biochem.*, **115**, 225–30 (1981).
10.17. J. Geigert, D. S. Hirano and S. L. Neidleman, *J. Chromatog.* **206**, 396–9 (1981).

10.18. A. W. Archer, V. R. Higgins and D. L. Perryman, *J. Assoc. Publ. Analysts*, **18**, 99–103 (1980).
10.19. M. Huong Bui-Ngúyén, *J. Chromatog.* **196**, 163–5 (1980). See also the review given by C. Hasselmann and P. A. Diop, *Sciences des Aliments*, 3, 161–80 (1983).

6
Vitamin C
(Ascorbic and Dehydroascorbic Acids) in Foodstuffs: Modified Deutsch and Weeks Fluorimetric Method

1. PURPOSE AND SCOPE

The method describes a procedure for the quantitative determination of vitamin C in foodstuffs. It determines the total vitamin C (ascorbic acid and dehydroascorbic acid) of natural origin and, if required, vitamin C added during the production of the foodstuff. The method is applicable to both fresh and stored foodstuffs intended for immediate consumption and in particular to those of complex composition, for example, vegetables, fruits, fruit juices, potatoes, complete meals, sausages, etc. Vitamin C contents above 1·5 mg per 100 g can be determined quantitatively.

2. DEFINITION

Vitamin C content is understood to be the ascorbic acid + dehydroascorbic acid content calculated as ascorbic acid determined by the procedure described here. It is given in mg of ascorbic acid per 100 g of original sample.

3. BRIEF DESCRIPTION (PRINCIPLE OF THE METHOD)

The sample is disintegrated in water–acetone–metaphosphoric acid solution, homogenised and extracted. Using aliquot parts of the filtered homogenate (=sample extract), the ascorbic acid is oxidised to dehydroascorbic acid by means of Norit activated carbon, the dehydroascorbic acid being derivatised to a fluorescent quinoxaline compound by a subsequent reaction with 1,2-phenylenediamine. Blocking of this reaction by boric acid, whereby a stable boric acid–dehydroascorbic acid complex is formed, provides the relevant blank value. A standard calibration curve is used for the calculations. Fluorescence is measured at 350/430 nm.

4. CHEMICALS

Remark: If not otherwise specified AR grade chemicals are to be employed. The water used must be distilled and taken from glass containers or be of corresponding purity.

4.1 Metaphosphoric acid, lumps, vitreous (40–44% (HPO_3), 56–60% ($NaPO_3$)), e.g. Riedel de Haën.
4.2. Boric acid, crystalline, H_3BO_3, e.g. Merck.
4.3 Sodium acetate trihydrate, highest purity p.a., >99·5%, e.g. Fluka or Merck (quality: no reaction to $KMnO_4$).
4.4 1,2-Phenylenediamine dihydrochloride, pure p.a., >99%, e.g. Fluka.
4.5. Norit activated carbon, neutral, $\Delta C-170$, Fisher Scientific Company.
    4.5.1. Acid-washed Norit: suspend 200 g Norit activated carbon (4.5) in 2000 ml HCl (4.9), bring to boil, cool to room temperature and evacuate on to a glass filter funnel (5.7). Suspend the filter cake in 2000 ml of water, filter off. Repeat the process several times (5–10) until the washing water appears neutral on pH paper. Dry the Norit overnight at 110–120 °C and pour into polyethylene bottles; it remains active for several months.

Note 1: The Norit carbon must be washed until it is completely free from acid. Hydrochloric acid residues greatly reduce its activity.

4.6. L (+)−Ascorbic acid (vitamin C) p.a. and for biochemical purposes, e.g. Merck. Store over P_2O_5, protected from light.
4.7. Acetic acid, 100% (at least 99·8%), e.g. Merck.
4.8. Hydrochloric acid, fuming, 37%, e.g. Merck.
4.9. Hydrochloric acid, approx. 3·7%; dilute 1 volume of HCl (4.8) with 9 volumes of water.
4.10. Acetone, highest purity, absolute and resistant to oxidation, e.g. Fluka.
4.11. Solutions.
    4.11.1. Solution A: metaphosphoric acid, 15% (w/v); 150 g metaphos-

phoric acid (4.1) dissolved in water to 1000 ml; filter off (5.7). Filtrate can be kept for up to 10 days at 4 °C.

4.11.2. Solution B: metaphosphoric acid, 7·5% (w/v); dilute 1 volume solution A (4.11.1) with 1 volume water; can be kept for up to three days at 4 °C.

4.11.3. Solution C: sodium acetate solution; 500 g sodium acetate trihydrate (4.3) dissolved in water to 100 ml; can be kept for one month.

4.11.4. Solution D: boric acid – sodium acetate solution; 3 g boric acid (4.2) dissolved in 100 ml solution C (4.11.3); to be freshly made up.

4.11.5. Solution E: extraction medium; 5 parts by weight of solution B (4.11.2) + 3 parts by weight of acetone (4.10); for example, 500 g solution B + 300 g acetone. Solution E can only be kept for one day! To be freshly made up.

Note 2: The extraction medium, solution E, has a different composition from that put forward by Deutsch and Weeks (Ref. 10.1); acetone is used instead of acetic acid. This avoids the inconvenient odour during the breaking down process. Acetone has three advantages: (1) it facilitates the filtering of the sample extract (7.2) by precipitating starch; (2) it dissolves the fats; (3) it accelerates the thawing out of deep-frozen sample extracts (7.2) because of the latter's semi-solid consistency at -20 °C which is due to the acetone.

4.11.6. Solution F: acetic acid – metaphosphoric acid solution; 300 ml solution A (4.11.1) + 80 ml acetic acid (4.7) made up with water to 1000 ml; to be freshly made up.

4.11.7. Solution G: 1,2-phenylenediamine dihydrochloride solution; 40 mg 1,2-phenylenediamine hydrochloride (4.4) dissolved in water to 100 ml. To be made up immediately before use!

4.11.8. Solution H: ascorbic acid standard solution; 25 mg L(+)-ascorbic acid (4.6), dried over P_2O_5, dissolved in solution F (4.11.6) to 100 ml; content: 0·25 mg ascorbic acid per ml.

5. APPARATUS AND AUXILIARY EQUIPMENT

5.1. Standard basic laboratory equipment.
5.2 Refrigerator.

5.3 Deep-freezer.
5.4. Drying cupboard.
5.5. Mixing apparatus, e.g. Waring blender; Polytron rod; oysters.
5.6. Wide-necked flask of elastic high-pressure polyethylene with screw fitting, 100 ml, e.g. Semadeni.
5.7. Glass filter funnel, 15–20 cm diameter.
5.8. Pleated filter, e.g. Schleicher and Schüll No. 595 1/2, 270 mm diameter.
5.9. Pleated filter, e.g. Schleicher and Schüll No. 595 1/2, 125 mm diameter.
5.10. Desiccator with P_2O_5 filling.
5.11. Commercially available fluorimeter; measuring cuvettes for 2 ml with complete cover of the measurement windows, excitation maxima at 350 nm, emission maxima at 430 nm.

6. SAMPLE

6.1. *Sampling*

The sample to be taken must be representative of the material to be investigated (foodstuffs) with regard to proportions and composition. It must in any case be taken from a sizeable initial amount.

6.2. *Sampling Technique and Preparation of the Sample*

Structured and inhomogeneous material should be mechanically reduced in size before the sample is taken for disintegration, homogenisation and extraction, to be carried out in a single operation (7.2). Where the precise vitamin C content is to be determined in moist and water-containing substances, a separate determination of its dry weight T_S (see equation in section 8.1.4) is necessary. To this end a given amount of the substance (g) with a given amount of water (g), normally in the ratio 1:1, is comminuted and homogenised. Approximately 15–20 g of this homogenate, weighed out exactly, is dried at 105 °C for 24 h or by any other standard process until a constant weight is obtained. The determination of T_S for precise calculation is not necessary up to 10% T_S.
Result: g T_S/100 g or T_S%. The dried substance is discarded.

6.2.1. *Weighed portion.* The weighed portions are normally 100 g of the foodstuffs.

7. PROCEDURE

7.1. Remark
The procedure is to be carried out without major interruptions. However, the sample extract (7.2) can be frozen at $-20\,°C$ to await further processing and can be stored in this state for one month. The total analysis is to be carried out as a true double determination (two weighed portions); all determinations should be carried out in parallel and should be performed in subdued light.

7.2. Disintegration and Extraction
Approximately 100 g of the water-containing sample prepared according to (6.2) and weighed to the nearest 0·1 g is diluted with twice its weight of the extraction medium, solution E (4.11.5), and further broken down in a mixer (5.5).

> *Note 3:* The weight ratio of water-containing sample to extraction medium solution E (in grams) must be maintained accurately at 1:2. Dry samples and samples with low water content can be diluted with 2–10 times the amount of extraction medium by weight so that a sample extract of approximately 80–100 ml can be obtained.

It is then further comminuted for 2 min with the Polytron rod (5.5) and homogenised; temperature should not rise (cool!). The homogenate is allowed to stand for 15 min and then filtered—without rinsing!—through a pleated filter (5.8). It is also possible to centrifuge the homogenate (take the upper phase). The filtered or centrifuged sample extract of approximately 80–100 ml is then either frozen (7.1) and stored or analysed immediately. Special polyethylene bottles (5.6) are suitable for the freezing process.

7.3. Ascorbic Acid Oxidation in the Sample Extract
A quantity (3·0 ml to a maximum of 25·0 ml) of the sample extract which may have been stored and thawed or freshly made—aliquots taken according to vitamin C content, containing at least 0·25 mg ascorbic acid—are made up to the mark in a 50 ml graduated flask with solution F (4.11.6). After pouring into a 250 ml ground-glass Erlenmeyer flask, 1 g of pretreated, acid-washed Norit (4.5.1) is added, the flask is shaken strongly for 15 s and the extract filtered through a pleated filter (5.9)—the first part of the filtrate is discarded (= sample filtrate P with total dehydroascorbic acid).

Note 4: The taking of aliquots is done volumetrically for simplicity; for the calculation (see equation in section 8.1.4) it is assumed that 1 ml ≃ 1 g, i.e. the density of the sample extract is on average about 1·0. Thawing is best carried out in cold water (10–13 °C) for approximately 1 h or at 4 °C overnight.

Note 5: If Norit is used in the non-pretreated, original commercial form (4.5), one has to wait and shake 30 min for a complete oxidation of the ascorbic acid, valid also for ascorbic acid standard solutions (7.4). See Ref. 10.3.

7.4. *Ascorbic Acid Oxidation in the Ascorbic Acid Standard Solution*

To plot a standard calibration curve (8.1.2), 1·0, 2·0, 3·0, 4·0 and 6·0 ml of solution H (4.11.8) corresponding to 0·25, 0·50, 0·75, 1·00 and 1·50 mg of ascorbic acid, respectively, are made up to the mark in a 50 ml graduated flask with solution F (4.11.6).

Note 6: The standard calibration curve must be set daily and then used for the whole day and for all types of sample.

Ascorbic acid standard concentrations of 5, 10, 15, 20 and 30 µg/ml result therefrom. These five solutions are poured into 250 ml ground-glass Erlenmeyer flasks and mixed with 1 g of pretreated, acid-washed Norit (4.5.1), shaken strongly for 15 s and filtered as in method 7.3., the initial filtrate being discarded (=five standard filtrates S with dehydroascorbic acid).

7.5. *Preparation of the Test Solutions*

7.5.1. *Sample test solution.* Five millilitres of sample filtrate P (7.3) are diluted with 5 ml of solution C (4.11.3) in a 50 ml graduated flask and made up to the mark with water (=sample test solution, P).

7.5.2. *Sample test solution—blank.* Five millilitres of sample filtrate P (7.3) are diluted with 5 ml of solution D (4.11.4) in a 50 ml graduated flask, allowed to stand for 15 min and shaken from time to time and then made up to the mark with water (=sample test solution—blank, P_B).

Note 7: The production of blanks for test solutions 7.5.2 and 7.5.4 can best be carried out in timed phases when series of samples are present: (a) take 5 ml of sample filtrate P (7.3) and standard filtrate S (7.4); (b) add 5 ml of solution D, accurately timed, after each 30 s; (c) allow to stand for

15 min; (d) make up to 50 ml, again accurately timed and equally ranged after each 30 s.

7.5.3. *Standard test solutions.* Five millilitres of each of the five standard filtrates S (7.4) are dealt with in the manner described in section 7.5.1 (= standard test solutions, S).

7.5.4. *Standard test solutions—blanks.* Five millilitres each of the five standard filtrates S (7.4) are treated in the manner described in section 7.5.2 (= standard test solutions—blanks, S_B). (See also Note 7.)

7.6. *Fluorimetry (Formation of Quinoxaline)*

A quantity (1·0 ml) of each of the test solutions P, P_B, S and S_B (7.5.1.– 7.5.4) with at least 0·5 μg ascorbic acid per ml are diluted in a test cuvette (5.11) with 1·0 ml of solution G (4.11.7), mixed and allowed to stand in the dark for 60 min. The emission at 430 nm with a 350 nm excitation is then measured (main and blank values).

Note 8: A reaction time of 60 min has been calculated as the optimum time; the solution begins to discolour if allowed to stand for any longer. The reading should be taken after exactly 60 min.

All the measurements are made in duplicate. The sample test values must lie within the limits of the calibration curves (8.1.2). This can be achieved by proper selection of the aliquots (7.3).

8. EVALUATION

8.1. *Calculation*

8.1.1. *Sample test value and standard test value.* These correspond to the net fluorescence as the difference between the mean values from two readings of the main and blank values (7.6).

8.1.2. *Standard calibration curve.* Standard test values (8.1.1) for the ascorbic acid standard concentrations (7.4) of 5, 10, 15, 20 and 30 μg of ascorbic acid per ml can be plotted or calculated by linear regression.

8.1.3. *Sample test values.* These values, in microgrames of ascorbic acid per ml, can be read off the standard calibration curve or calculated.

8.1.4. *Calculation of the ascorbic acid content.* The ascorbic acid content in 100 g of the sample material investigated can be calculated by the equation:

$$\frac{\text{mg ascorbic acid}/100 \text{ g}}{\text{original sample}} = \frac{a \times 50 \times (c - T_S) \times 100}{b \times E \times 1000}$$

$$= \frac{a \times 5 \times (c - T_S)}{b \times E}$$

where: a = sample test value for total ascorbic acid read off the calibration curve (8.1.2) or calculated (8.1.3) (µg/ml); b = quantity (ml) in the aliquot (7.3) (see also Notes 4 and 5) (g); c = total weight of the homogenate (7.2) (weighed portion + twice the amount of extraction medium) (g); T_S = dried substance (6.2) of the weighed portion (g); E = weighed portion (6.2.1) (g); 50 = aliquot made up to 50 ml (7.3); 100 = conversion to 100 g sample; 1000 = conversion to mg ascorbic acid.

8.1.5 *Range of linearity.* The linearity of the calibration curve is in the range 0·5–5 µg ascorbic acid per ml of standard test solution.

8.1.6. *Detection limit.* An amount of 1·5 mg of ascorbic acid per 100 g of sample or 0·5 µg per ml of standard test solution can be quantitatively evaluated. The upper detection limit for this method is 250 mg per 100 g of sample.

8.2. *Reliability of the Method*

8.2.1. *Recovery.* For sausages a recovery rate of 97% was found in the supplementary test; no values are available for vegetables.

8.2.2. *Repeatability.* This was not determined.

8.2.3. *Comparability.* This was not determined (no syndicate tests).

8.2.4. *Comparison with other methods for determining vitamin C.* No comparisons are available.

9. ANALYSIS REPORT

The result of the determination is to be given with a reference to this method. Operations not mentioned in the described method have to be indicated.

10. REFERENCES

10.1. M. J. Deutsch and C. E. Weeks, *J. Assoc. Off. Anal. Chem.*, **48** (6), 1248–56 (1965).
10.2. P. Scheffeldt, Institut für Ernährungsforschung, Seestrasse 72, CH-8803 Rüschlikon–Zürich.
10.3. D. Scuffam and J. R. Cooke, Laboratory of the Government Chemist, London. Private communication, 1983.

7

Vitamin C (Ascorbic and Dehydroascorbic Acids) in Food: Sephadex Method

1. PURPOSE AND SCOPE

The method describes a procedure for the quantitative determination of total vitamin C content in food which includes naturally occurring vitamin C (ascorbic acid + dehydroascorbic acid) in the constituents of the food and ascorbic acid which may be added during its manufacture. The method may be applied to fresh and stored foods intended for immediate consumption, e.g. all kinds of vegetables, cereals, dairy and meat products, fish, fruits, desserts, beverages and ready-cooked meals. No foods have been identified for which the method is not applicable. Quantitative determination of vitamin C content is possible down to 0·5 mg ascorbic acid in 100 g of food.

2. DEFINITION

Vitamin C content is taken to mean the sum of ascorbic and dehydroascorbic acids determined by this method. Values are given as mg ascorbic acid per 100 g of sample.

3. BRIEF DESCRIPTION (PRINCIPLE OF THE METHOD)

After disintegration, homogenisation and extraction of the material under investigation in aqueous metaphosphoric acid, the extract is treated with 2,3-dimercapto-1-propanol (BAL) to reduce dehydroascorbic acid (DASC) and to eliminate metal ions. Excess of BAL is removed by reaction with N-ethylmaleimide (NEM), and the resulting total ascorbic acid (ASC) separated by percolating the treated extract through a column of anionic Sephadex. After washing with water, ASC is oxidised on the column to DASC by a *p*-benzoquinone solution, which simultaneously elutes the DASC, which has no acidic properties and is therefore no longer retained. The eluate, containing DASC, is treated with 4-nitro-1,2-phenylenediamine (NPD); the yellow reaction product is colorimetrically measured at 375 nm.

4. CHEMICALS

Remark: Unless otherwise specified reagent-grade chemicals are to be used: water must be either double distilled or of equivalent purity.

4.1. Metaphosphoric acid $(HPO_3)_n$, purum, sticks, approx. 65% (HPO_3), approx. 35% $(NaPO_3)$; Fluka No. 79615 or Prolabo No. 20632 or equivalent product of Carlo Erba. Do not use acid with $(NaPO_3)$ content of 56–60%.

4.2 Potassium hydroxide, pellets, min. 85%; Merck No. 5033.

4.3 Sephadex ion exchanger DEAE A-25 (chloride form); Pharmacia Fine Chemicals, Code No. 17-0170-01 and 17-0170-02.

4.4 Sephadex G-25 coarse; Pharmacia Fine Chemicals, Code Nos 17 0034-01 and 17-0034-02.

4.5. *p*-Benzoquinone (*p*-quinone; 1,4-benzoquinone), puriss. >99·5%; Fluka No. 12309, Prolabo No. 27346 (must be yellow!).

4.6. Dipotassium hydrogen phosphate trihydrate $K_2HPO_4 \cdot 3H_2O$, min. 99%; Merck No. 5099.

4.7. 4-Nitro-1,2-phenylenediamine (NPD), p.a.; Merck No. 11657.

4.8. Sodium acetate trihydrate, puriss. p.a. >99·5%; Fluka No. 71190 or Merck No. 6267 or Prolabo No. 27652.

4.9. L (+)-Ascorbic acid (vitamin C), p.a. and for biochemical purposes; Merck No. 127. Store light-protected over P_2O_5.

4.10. Sodium hydroxide, pellets, min. 99%; Merck No. 6498.

4.11. Silver nitrate, min. 99·8%; Merck No. 1512 or Prolabo No. 21572.

4.12 Aluminium oxide, neutral, activity I; Woelm.

4.13. *N*-Ethylmaleimide (NEM), puriss. *ca.* 99%; Fluka No. 04260.

4.14. Orthophosphoric acid, 85%; Merck No. 573 or Prolabo No. 20624.

4.15. Acetic acid, glacial, 100%, min. 99·8%; Merck No. 63 E.

4.16. 2,3-Dimercapto-1-propanol for complexometry (British Anti-Lewisite = BAL); Merck No. 3409.

4.17. Chloroform, stabilised with ethanol; Merck No. 2445. Saturate 1 litre with 200 ml water and leave in contact with water. Discard after one week.

4.18. Tetrahydrofuran (THF), stabilised with BHT; Merck No. 9731. If peroxides are present they should be removed by percolating 50 ml through a Woelm Al_2O_3 (4.12) column made of Allihn tube (5.5). The THF must be free of the stabiliser hydroquinone.

4.19. Ethyl acetate; Merck No. 9623. Saturate 1 litre with 200 ml water by stirring for 1 min; prepare 2 h before use and leave in contact with water. Keep no longer than one week at 20 °C.

4.20. Nitrogen (free of oxygen, 99·9% (v/v)).

4.21. Solutions.

    4.21.1 Metaphosphoric acid, 20% (w/v) aqueous solution of metaphosphoric acid (4.1). Wash sticks with water, weigh 200 g of the acid (4.1) and adjust to 1 litre with water. Do not heat, but shake from time to time and filter the solution; it must be kept in the refrigerator for not more than one week.

    4.21.2. Metaphosphoric acid, 5% (w/v) aqueous solution of metaphosphoric acid (4.1). Prepare each day from acid (4.21.1) by dilution.

    4.21.3 Potassium hydroxide, 50% (w/v) aqueous solution. Weigh 29·5 g KOH (4.2) and adjust to 50 ml with water. Keep for one week.

    4.21.4. Acetic acid, 10% (v/v) aqueous solution from glacial acetic acid (4.15).

    4.21.5. p-Benzoquinone, 0·05% (w/v) aqueous solution. Weigh 100 mg of p-benzoquinone (4.5) and adjust to 200 ml with water. Must be kept in darkness; keep for no longer than one day.

    4.21.6. Orthophosphoric acid, 18·4% (w/v) aqueous solution. Weigh precisely 21·6 g of orthophosphoric acid (4.14) and adjust to 100 ml with water. Keep for one week.

    4.21.7. Di-potassium hydrogen phosphate, 32% (w/v) aqueous solution. Weigh precisely 42·0 g of the salt (4.6) and adjust to 100 ml with water. Keep for one week.

    4.21.8. 4-Nitro-1,2-phenylenediamine, 0·6% (w/v) solution in tetrahydrofuran. Weigh 600 mg NPD (4.7) and adjust to 100 ml with THF (4.18). Prepare freshly with purified THF.

    4.21.9. Sodium acetate, aqueous solution. Weigh 100 g of sodium acetate (4.8) and add 100 ml water. Keep for one week.

    4.21.10. Sodium hydroxide, about 48% (w/v) aqueous solution. Weigh 48·0 g NaOH (4.10) and adjust to 100 ml with water. Keep for one week.

4.21.11. Sodium hydroxide, 0·5 M (0·5 N) solution; Merck No. 9138.
4.21.12. Basic buffer; mix 35 ml sodium acetate (4.21.9) and 15 ml NaOH (4.21.10). Prepare freshly immediately before use.
4.21.13. *N*-Ethylmaleimide, 3·2% (w/v) aqueous solution. Dissolve 3·2 g NEM (4.13) in 80 ml water at 40 °C, cool and adjust to 100 ml with water. Keep cool at +5 °C in a refrigerator.
4.22. Preparation of Sephadex DEAE A-25: weigh 100 g Sephadex DEAE A-25 (4.3) and add 2 litres of water. Filter through a glass frit (5.6). Allow NaOH (4.21.11) to flow through the Sephadex until the eluate is free of chloride ions (collect 2 ml eluate, acidify to about pH 3 with a few drops of acetic acid (4.21.4), add a few drops of a solution of silver nitrate (4.11): a precipitate is formed when chloride ions are present)—quantity of NaOH (4.21.11) used, about 8 litres; duration, about 1–2 h. Then add water immediately and rinse until the water is neutral (pH 5–6)—quantity of water used, about 10–20 litres; duration, about 1–2 h. Store in the refrigerator under water for a maximum period of one month. It may be necessary to change the water if it becomes basic again (use about 500 ml water for 100 g Sephadex).
4.23. Suspension of Sephadex G-25 coarse: aqueous slurry of (4.4); keep in the refrigerator for several weeks; see section 7.4.1.
4.24. Standard solutions of ascorbic acid: prepare solutions which contain 8, 16, 24 and 32 µg ascorbic acid (4.9) per g metaphosphoric acid, 5% (w/v) (4.21.2). These solutions can easily be prepared by dilution of a solution of 128 mg ascorbic acid (4.9) in 100 g metaphosphoric acid, 5% (w/v) (4.21.2) with this acid.

5. APPARATUS AND AUXILIARY EQUIPMENT

5.1. Standard laboratory basic equipment.
5.2. Rotary evaporator with water bath.
5.3. Water bath with thermostat.
5.4. Apparatus for homogenisation of food samples; see section 6.2: mixer, Waring blender; preferably Polytron^R
5.5. Allihn tube, 10 × 2 cm.
5.6. Glass frit No. 3, diameter 17·5 cm.
5.7. Glass chromatography column (see Fig. 12), two pieces.
5.8. Folded paper filter.
5.9. Centrifuge.

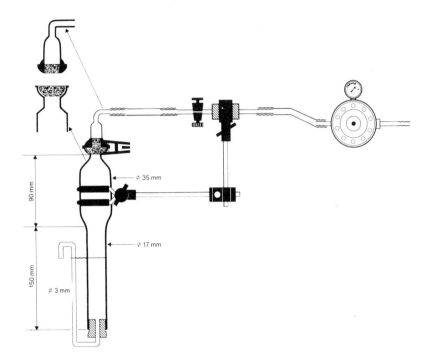

Fig. 12. Chromatography column with manometer for ascorbic acid determination.

5.10. Refrigerator.
5.11. pH meter.
5.12. Spectrophotometer.
5.13. Glass cuvettes, 1 cm.
5.14. Manometer for nitrogen (see Fig, 12)

6. SAMPLE

6.1. *Sampling*
The sample must be representative as regards proportion and composition of the material under investigation. A large sample should always be taken (250–500 g).

6.2. Sampling Method and Preparation of the Sample

Structured and inhomogeneous material must be mechanically comminuted to small pieces before the sample is taken for disintegration, homogenisation and extraction, carried out in a single operation in a suitable mixer (5.4). It must be analysed immediately.

6.2.1. Weighed portion. The weighed portion should preferably be in the range of 10–20 g and not greater than 50 g because of the size of the equipment used.

7. PROCEDURE

7.1. Remark

Two replicate samples should be analysed (two weighed portions (6.2.1)).

7.2. Disintegration, Homogenisation and Extraction

Weigh 10 or 20 g of the sample and add it to 90 or 80 g of the HPO_3 solution (4.21.2). Mix the suspension using a Waring blender (5.4) for 1 min and centrifuge at 3000 rpm for 15 min. Filter through a folded paper filter (5.8) (=extract).

> *Note 1:* If ASC alone is to be determined, take an aliquot of 20 g of the extract (7.2), dilute with 5% metaphosphoric acid (4.21.2) to 100 ml; 10·0 ml of this solution is put onto the column according to step 7.4.2 whilst omitting the preceeding step 7.3. The extract (7.2) can be kept for one night in the refrigerator at this point.

7.3. Reduction of DASC

7.3.1. Extracts (7.2) containing less than 25 µg ASC per g extract:
(a) To 30 g of the extract (7.2) add 0·1 ml BAL (4.16), stir and adjust the pH to 5·5 with 50% KOH solution (4.21.3).
(b) Allow to react for 10 min at 30 °C (water bath (5.3)). The solution can be stored for one night in the refrigerator at this point.
(c) Add 10 ml of the NEM solution (4.21.13) and allow to react at 30 °C for precisely 1 min (inactivation of excess BAL).
(d) Acidify immediately with 5 ml of 20% HPO_3 (4.21.1) and adjust the volume to 50 ml with water (=reduced extract; it contains total vitamin C in the form of ASC).

> *Note 2:* It is possible to take 60 g of the extract and to double the quantities of reagents and adjust to 100 ml (7.3.1(d)) in the case of special prepurification. To avoid corrosion of the electrodes of the pH meter

(5.11) by BAL wash them with ethanol (never keep for longer than 30 min in ethanol).

7.3.2. *Extracts (7.2) containing more than* 18 µg *ASC per g extract:* following steps 7.3.1 (a)–(d), refer to Table 3 for volumes and dilutions (=reduced extract).

7.4. *Chromatographic Purification and Oxidation*

7.4.1. *Preparation of the columns.* Take sufficient refrigerated suspension of prepared Sephadex DEAE A-25 (4.22) for 1 day. Filter through glass frit No. 3 (5.6) under vacuum to eliminate water and place the Sephadex into a 10% acetic acid solution (4.21.4) in order to obtain the acetate form from the free amine form of the ion exchanger (250 ml acetic acid (4.21.4) for 50 g Sephadex (4.22)). Stand for one hour at room temperature. Fill the columns (5.7) with the slurry and apply a pressure of nitrogen (4.20) of approx. 0·6 bar. On top of the Sephadex DEAE A-25 put an aqueous slurry of Sephadex G 25 (4.23), about 1 cm high. Operate in subdued light so as to avoid UV radiation.

Note 3: Sephadex G-25 coarse is neutral; it helps to regulate the flow and protects the filling.

7.4.2. *Purification and oxidising elution.* Percolate slowly 10·0 ml of the reduced extract (7.3.1 (d) or 7.3.2 (d)) containing between 50 and 200 µg ASC until the top level of the Sephadex DEAE A-25 (not G-25!) is almost dry. Do not let the Sephadex DEAE dry out! This operation should not take less than 2 min (it lasts from 2 to 4 min); the pressure must then be reduced to about 0·06 bar and the column left in contact with the reduced extract for 10 min.

Rinse the columns with 50 ml water *in one run* and in less than 5 min under a pressure of about 0·6 bar (this is generally enough to eliminate impurities); in some cases it will be necessary to repeat this operation. Add 15 ml *p*-benzoquinone solution (4.21.5) on top of the column and percolate slowly (minimum duration 4 min; in general this takes 4–7 min). Collect the eluate in a 20 ml volumetric flask and adjust to volume with water; transfer immediately to a separating funnel and extract twice with 50 ml water-saturated chloroform (4.17) to remove the excess of *p*-benzoquinone. Shake each time for 15 s. The aqueous layer contains DASC (=DASC solution). It is possible to keep the DASC solution at this point in the refrigerator at +4 °C for up to 3 h.

TABLE 3
Determination of Total Vitamin C (ASC+DASC). Reduction of the Extract with BAL and Inactivation of Excess BAL with NEM

	Conditions needed for concentrations of total vitamin C in the extract (7.2) of:		
	8–25 µg/g	18–36 µg/g	30–60 µg/g
Weight of the extract to be treated (g)	30	25	25
Volume of BAL (4.16) (ml)	0·1	0·1	0·1
Adjust with KOH solution (4.21.3) to pH (pH units)	5·5	5·5	5·5
Reaction time at 30 °C (reduction) (min)	10	10	10
Volume of NEM solution (4.21.13) (ml)	10	10	10
Reaction time at 30 °C (inactivation) (min)	1	1	1
Volume of 20% HPO$_3$ solution (4.21.1) to be added (ml)	5	6·3	6·3
Adjust volume with water to (ml)	50	50	50
Dilution of the reduced extract (7.3.1 (d)) or (7.3.2 (d)) with 5% HPO$_3$ solution (4.21.2)	No	No	15 ml aliquot adjusted to 25 ml
Volume layered on the column (ml)	10	10	10
Quantity of vitamin C layered on the column. Concentration limits (µg)	56–175	90–180	90–180

Note 4: (1) The loss of ASC is nil with a 50 ml washing and negligible up to 200 ml; however, it is preferable to check the loss when more than two washings are made (i.e. more than 100 ml). Numerous washings with portions less than 50 ml are to be avoided since ASC is a weak acid and has a tendency to leave the Sephadex because of the pressure shock on the column. Very slow washings are to be avoided for the same reason. If a column is overloaded too much to be washed with 100 ml, then it is preferable to purify the sample extract before the chromatographic separation (see (3)). If the ASC solution contains too much mineral salt, it is possible to increase the column height, but it is then necessary to increase the volume of *p*-benzoquinone solution.

(2) Liquids poured onto the columns must be at the same temperature as the Sephadex.

(3) Starch and (or) dextrin containing products: weigh 60 g of the extract (7.2), add 60 g ethanol, agitate and centrifuge. Filter the liquid through a folded paper filter, collect 80 g of the filtrate and evaporate the ethanol completely under vacuum at 40 °C with a rotary evaporator (5.2). Adjust the weight of the residue to precisely 40 g with double distilled water. When the product contains dextrin replace the ethanol by methanol and use a larger amount. It will be applicable to potatoes (starch) and nutrients for children (dextrin)—infant formulas.

(4) The metaphosphoric acid solution layered on the Sephadex columns must be $5 \pm 1\%$ HPO_3.

7.5. *Colorimetry*

7.5.1. *Buffer control.*
To 20 ml of water add 4·0 ml of H_3PO_4 (4.21.6) and 4·0 ml of K_2HPO_4 (4.21.7). The pH must be between 3·8 and 4·0.

Note 5: If the pH lies outside this range modify the H_3PO_4 concentration accordingly. A variation of 0·1 unit of pH corresponds to about 270 mg of 85% H_3PO_4 (4.14) in 100 ml of the solution (4.21.6).

7.5.2. *Derivatisation of DASC with NPD.*
Bring the temperature of the DASC solution (7.4.2) to +30 °C and buffer it with at first 4·0 ml H_3PO_4 (4.21.6), then 4·0 ml K_2HPO_4 (4.21.7), and finally 2·0 ml of the NPD reagent (4.21.8). An opalescence is to be seen when adding this reagent due to the presence of traces of chloroform; it disappears after a few minutes. Leave for 75 min at 30 °C in a water bath (one can leave the solutions at 30 °C up to 3 h).

7.5.3. *Elimination of excess NPD.*
Add 5·0 ml of basic buffer (4.21.12). Check that the pH is about 10·8. If it lies outside 10·5–11·0 adjust the concentration of solution (4.21.10) accordingly. Extract the excess NPD,

once with 100 ml water-saturated ethyl acetate (4.19) and once with 50 ml of the same solvent (shake vigorously for 15 s each time). Collect the aqueous layer and strip off the dissolved ethyl acetate by bubbling nitrogen (4.20) for 1 min into the solution. It is not possible to eliminate ethyl acetate completely, and it is not necessary to adjust the volume of the solution (=sample test solution).

7.5.4. *Measurement.* In a time interval not exceeding 7 min after addition of the basic buffer, read the optical absorption of the sample test solution (7.5.3) at 375 nm in a 1 cm glass cuvette (5.13) against water.

7.5.4.1. Blank for chemicals (reagents) control: Pour 10 ml of metaphosphoric acid (4.21.2), containing no ASC, on top of a Sephadex column and proceed as for a normal determination (sections 7.4.2–7.5.4). The optical absorption should not exceed 0·030.

8. EVALUATION

8.1. *Calculation*

8.1.1. *Standard curve.* Standard solutions of all concentrations (4.24), each of 30 g, are analysed according to steps 7.3.1 and 7.4.2–7.5.4. Extinction values are obtained of the corresponding standard test solutions (analogously to section 7.5.3). They form a straight line of slope s, i.e.

$$y = sx + a \tag{1}$$

where a = blank for chemicals (7.5.4.1), y = optical absorption found at 375 nm, x = concentration of ASC in the standard solutions (4.24) and s = slope of the curve.

8.1.2. *Content of ASC in the sample.* The content of ASC in the material under investigation may be calculated by the concentration of ASC found in the extract (7.3.1) using the equation:

$$[\text{ASC}] = \frac{y-a}{s} \tag{2}$$

and with regard to the weighed portion (7.2) and aliquots taken during the procedure steps 7.3.1 or 7.3.2 and 7.4.2. It is equally possible to read the ASC values directly from a drawn curve without using eqn (1).

8.1.3. *Range of linearity.* The linearity of the regression line was highly significant at a level of 0·1%. The coefficient of variation of the slope, obtained from 24 standard curves each made with four points, in an interval of one month, was only 2·4%.

8.1.4. *Detection limit.* The detection limit depends on the nature of the sample; the lower limit is about 0·5 mg ASC per 100 g of the sample. Quantitative detection is possible down to 0·8 µg per ml of the test solution (7.5.3).

8.2. Reliability of the Method

Recovery was found in the range of 95–103% when ASC was added to the homogenised samples (6.2). Twelve different samples gave an average recovery of 99·3% (each sample was analysed once). In seven years more than 3000 food samples have been analysed by the authors; it is not necessary to introduce a recovery correction to the calculation.

8.2.2. *Repeatability.* The repeatability in the range of 50–200 µg ASC per 10 g standard solution (4.24) corresponds to a coefficient of variation of about ±5%. In detail: for 50 µg, 6·2%; 100 µg, 5·3%; 150 µg, 4·1%; 200 µg, 3·2%.

8.2.3. *Comparability.* No syndicate tests have been made.

8.2.4. *Comparison with other vitamin C determination methods.* The described method was compared with three other independent methods (see Ref. 10.5): *o*-phenylenediamine method, dinitrophenylhydrazone method and HPLC method, when ASC content was determined in 16 samples of different fresh vegetables and fruits. The results obtained showed, in general, very good agreement (see Ref. 10.5, p. 196, Table 1).

9. ANALYSIS REPORT

The result of the determination is to be given with a reference to this method. Operations not mentioned in the described method have to be indicated.

10. REFERENCES

10.1. S. Ogawa, *J. Pharmacol. Soc. Jap.* **73**, 59 (1953).
10.2. E. Gero and A. Candido, *Intern. Z. Vitaminforsch.*, **39**, 252 (1969).
10.3. C. F. Bourgeois and P. R. Mainguy, *Analusis* **2** (8), 556–61 (Sept. 1973).
10.4. C. F. Bourgeois, A. M. Czornomaz, P. George, J.-P. Belliot, P. R. Mainguy and B. Watier, *Analusis*, **3** (10), 540–8 (1975).
10.5. A. Floridi, R. Coli, A. Alberti Fidanza, C. F. Bourgeois and R. A. Wiggins, *Intern J. Vit. Nutr. Res.*, **52**, 194–7 (1982).
10.6 The described method has been successfully used over several years at:

1. Institut Scientifique d'Hygiène Alimentaire à Champlan auprès de Paris (Directeur: Dr Luquet), France.
2. Centre de Recherche Foch à Paris (Directeurs: M. Gounelle de Pontanel and Mme Astier-Dumas), France.
3. F. Hoffmann-La Roche & Co., Laboratoire BTA, Fontenay-sous-Bois, among other institutes in France.

8
Vitamin E (Only α-Tocopherol) in Foodstuffs: HPLC Method

1. PURPOSE AND SCOPE

The method describes a procedure for the quantitative determination of vitamin E (α-tocopherol) in foodstuffs. It determines the α-tocopherol of natural origin and, if required, α-tocopherol in free or esterified form added during the production of the foodstuff. The method is applicable to both fresh and stored foodstuffs of all kinds intended for immediate consumption. The quantitative determination and evaluation of the three other tocopherol homologues (β-, γ- and δ-) and of the four corresponding tocotrienols are not covered by this determination method (for structural formulae, see Fig. 8). α-Tocopherol contents above 0·1 mg/100 g of sample can be determined quantitatively, and contents of 0·05 mg/100 g can be detected qualitatively.

2. DEFINITION

The content of vitamin E is understood to be α-tocopherol content determined by the procedure described here. It is given in mg of α-tocopherol/100 g of sample.

3. BRIEF DESCRIPTION (PRINCIPLE OF THE METHOD)

After alkaline saponification of the sample (elimination of fats, release of natural tocopherols from the cellular tissue, hydrolysis of tocopherol esters to free tocopherols), and exhaustive extraction of the non-saponifiables with diethyl ether, the extract residue is dissolved in n-hexane. The α-tocopherol is measured fluorimetrically after separation on a straight phase HPLC column.

Note 1: The straight phase HPLC method described here enables all eight tocopherols and tocotrienols to be separated. The order in which they appear from the column is as follows: α-T, α-T$_3$, β-T, γ-T, β-T$_3$, γ-T$_3$, δ-T and δ-T$_3$, with retinol some way behind (see Fig. 14). Only the α-tocopherol signal is evaluated here. In contrast to older GC methods the α-tocopherol is fully separated from α-tocotrienol which is of importance in, for example, the determination of vitamin E in cereals. The α-tocopherol tables therefore are now outdated and require revision (see Fig. 14(c)).

4. CHEMICALS

Remark: Unless otherwise specified, AR grade chemicals are to be used. Distilled water in glass vessels or water of corresponding purity must be used.

4.1. Potassium hydroxide, pellets, extremely pure (85%), e.g. Merck.
4.2. L (+)−Ascorbic acid (vitamin C) AR grade and for biochemical purposes, e. g. Merck.
4.3. all-*rac*-α-Tocopherol for biochemical purposes, e.g. Merck.
4.4. BHT (2,6-di-*tert*-butyl-4-methylphenol) for synthesis, e.g. Merck.
4.5. Diethyl ether, AR grade, stabilised with BHT, e.g. Merck.
4.6. *n*-Hexane for fluorescence spectroscopy, Uvasol^R, e.g. Merck.
4.7. 1,4-Dioxane for spectroscopy, Uvasol^R, stabilised with BHT, e.g. Merck.
4.8. Methanol, e.g. Merck.
4.9. Ethanol, absolute, e.g. Merck.
4.10. Nitrogen, oxygen-free, 99·9% (v/v).
4.11. Solutions.
    4.11.1. Ascorbic acid solution 0·5%(w/v); 0·5 g ascorbic acid (4.2) are dissolved in 4 ml water and made up to 100 ml with methanol (4.8).
    4.11.2. Potassium hydroxide solution, 50% (w/v) aqueous; 50 g KOH (4.1) are dissolved in water with cooling to make 100 ml.
    4.11.3. α-Tocopherol standard solution. Approx. 50 mg all-*rac*-α-tocopherol (4.3) weighed to an accuracy of 0·01 mg are dissolved in *n*-hexane (4.6) with the addition of a few crystals of BHT (4.4) and made up to 100 ml (=stock solution with approx. 0·5 mg α-tocopherol/ml). The solution can be kept for up to six months at +4 °C. A quantity (2·0 ml) of the stock solution is diluted with *n*-hexane (4.6) to make 100 ml (=α-tocopherol standard solution with approx. 10 µg α-tocopherol/ml). The solution can be kept for up to two weeks at +4 °C.

5. APPARATUS AND AUXILIARY EQUIPMENT

5.1. Standard basic laboratory equipment including refrigerator and deep freezer.
5.2. Ground-glass round flask (250 ml) with reflux condenser and N_2 supply in the condenser.
5.3. Conical separating funnel, Squibb, 500 ml.
5.4. Mixing apparatus, e.g. Waring blender, Polytron rod, etc.
5.5. Coffee grinder.
5.6. Water bath or steam bath; heating mantle.
5.7. Rotary evaporator with water bath.
5.8. HPLC device.
   Normal commercially available apparatus, Perkin–Elmer, Waters, etc. Integrator, e.g. Spectra Physics 4100; Varian CDS 111 Recorder, commercially available.
   Spectrofluorimeter, e.g. Perkin–Elmer 3000; 293/326 nm with xenon flash tube (8 W) pulsed xenon lamp.
   Stationary phase: RT ready-made column, LiChrosorb Si 60, 5 µm, 125 × 4 mm, e.g. Merck (Hibar).

6. SAMPLE

6.1. *Sampling*
The sample to be taken must be representative of the foodstuffs to be investigated with regard to proportions and composition. In any case a sizeable initial quantity (100–250 g) must be available.

6.2. *Sampling method and preparation of the sample*
Solid samples should be ground down (5.5) to a fine consistency and thoroughly mixed; heating of the sample should be avoided. Pulpy foods with a higher water content are best reduced and homogenised in a mixer (5.4); BHT should be added as a protection against oxidation if vitamin A is to be determined at the same time. The prepared samples should be further processed immediately or stored in the deep-frozen state in a freezer.

   Note 2: Deep-frozen prepared samples can be kept for up to three months without any α-tocopherol losses.

6.2.1. *Weighed portion.* For α-tocopherol contents between 0·1 and 50 mg per 100 g of sample, weighed portions of 10 g are prepared; for oils, butter and fats a maximum of 2 g will suffice.

7. PROCEDURE

7.1. Remark
The procedure is to be carried out without major interruption as a true double determination (with two weighed samples). Direct sunlight should be avoided. Possible interruption points are indicated.

7.2. Saponification
Approx. 10 g of the sample prepared according to method 6.2, weighed accurately to 0·01 g, are poured into a 250 ml round flask (5.2) with reflux condenser with 50 ml of the methanolic ascorbic acid solution (4.11.1). After the addition of 10 ml of potassium hydroxide solution (4.11.2) the nitrogen (4.10) is drawn through the condenser, the air expelled, and the mixture is heated to boiling on the water or steam bath or with the heating apparatus (5.6). Simmering is continued for 20 min with suitable agitation, and afterwards the mixture is cooled down to 30 °C. The still fairly warm contents of the flask are transferred with approximately 50 ml water in two to three portions into a 500 ml conical separating funnel (5.3) containing 100 ml diethyl ether (4.5) (=saponification solution).

7.3. Extraction
After the conical funnel has been vented several times by opening the tap, the saponification solution (7.2) is shaken vigorously for 40 s and extracted. After phase separation the extraction is repeated twice with 100 ml diethyl ether (4.5). The combined extracts are washed with water until free of alkalis; in general three washes with 100–120 ml water each time will suffice (phenolphthalein control or pH paper (pH 7)) (=extract).

Note 3: The extract should be further treated immediately (7.4).

7.4. Sample Test Solution
The extract (7.3) is carefully evaporated to dryness in the rotary evaporator (5.7) under a partial vacuum at a maximum of 40 °C water bath temperature; to remove the rest of the water 10 ml of *n*-hexane

(4.6) + 10 ml absolute ethanol (4.9) are added and the evaporation process is repeated. The dry residue is dissolved in 20·0 ml *n*-hexane (4.6) (= sample test solution).

> *Note 4:* With high α-tocopherol contents 50·0 ml *n*-hexane (4.6) are used, while 10·0 ml are used with low contents; 20·0 ml *n*-hexane have proved sufficient for food analysis. The sample test solution should be adjusted for 5–10 µg α-tocopherol/ml. It can be kept overnight under nitrogen at +4 °C without any α-tocopherol losses. If retinol is to be determined in the test solution, approx. 5 mg BHT (4.4) should be added against oxidation to the extract (7.3) before evaporation to dryness.

The flask should always be opened under nitrogen (4.10) to avoid oxidation of solvent-free α-tocopherol.

7.5. *Standard Test Solution*

The α-tocopherol standard solution (4.11.3) containing 10 µg all-*rac*-α-tocopherol (4.3)/ml is also the standard test solution. It can be used for the external calibration of the HPLC line and for checking the stability (consistency) of the spectrofluorimeter; it is not subjected to full analysis.

7.6. *HPLC*

7.6.1. *Calibration of the HPLC line.* Before a measurement series is started, the HPLC line is calibrated with a double measurement using 50 µl of the standard test solution (7.5), which corresponds to 0·5 µg (500 ng) of α-tocopherol (mean value). The signal is evaluated either by means of peak height (chromatogram) or peak area (integrator). The chromatography process is carried out isocratically at room temperature. The calibration should be checked repeatedly during the series of measurements (every fifth injection).

7.6.2. *Measurement of the samples.* For every 50 µl of the sample test solution (7.4), two injections are made as described in section 7.6.1 and the peak signal evaluated accordingly (mean value).

7.6.3. *Chromatographic conditions:*
Apparatus (5.8)
Stationary phase: LiChrosorb Si 60, 5 µm.
Capacity: 50 µl (by loop).
Mobile phase: 3% dioxane (4.7) in *n*-hexane (4.6) (v/v).

Pressure: Approx. 30 bars.
Flow rate: 1 ml/min.
Detection: Fluorimetry 293/326 nm and with the retention time determined for α-tocopherol during calibration (7.6.1).
Analysis time: 30 min.
Peak evaluation: Either by height or integrated area.

Note 5: The 30 min analysis duration is the result of the delayed elution of the retinol from the column. If retinol is present, it must be waited for because it absorbs light at 326 nm and would cause quenching of the α-tocopherol signal.

The frequent practice of removing the retinol with an 80% sulphuric acid washing of the hexane-absorbed extract cannot be used in fluorescence detection, since any α-tocotrienol which may be present will be converted into three derivatives with relative retention times of 0·95, 1·00 and 1·14—taking α-tocopherol to be 1·00—and thus provide misleading results for α-tocopherol.

7.6.4. *Chromatograms.* Figures 13 and 14 show HPLC chromatograms of the determination of α-tocopherol in the α-tocopherol standard solution (4.11.3) and in mixed tocopherols, and in two complete meals, as well as in barley grain. The tocopherol homologues are also marked where present.

8. EVALUATION

8.1. Calculation

The peak heights or areas of the sample test solutions (7.4) are compared with those of the standard test solution (7.5); weighing, aliquots and dilutions should be considered thereby.

8.1.1. *Range of linearity.* The linear proportionality of the peak heights or areas for the sample test solutions as against those of the standard (500 ng) is in the range 1–50 µg α-tocopherol/ml, corresponding to 20–1000 ng per injection (50 µl).

8.1.2. *Detection limit.* An amount of 100 µg α-tocopherol in 100 g sample or 1 µg/ml of sample test solution can be quantitatively evaluated; 50 µg in 100 g can be determined qualitatively.

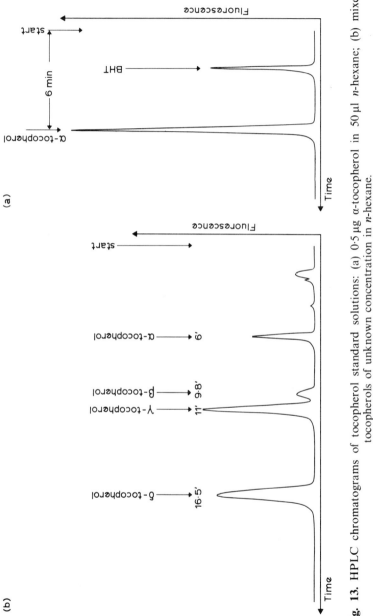

Fig. 13. HPLC chromatograms of tocopherol standard solutions: (a) 0·5 μg α-tocopherol in 50 μl n-hexane; (b) mixed tocopherols of unknown concentration in n-hexane.

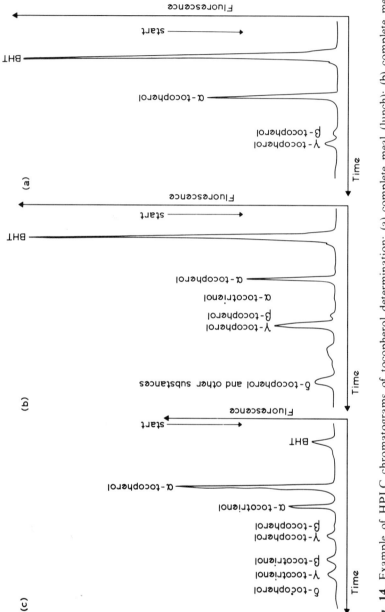

Fig. 14. Example of HPLC chromatograms of tocopherol determination: (a) complete meal (lunch); (b) complete meal (dinner); (c) crude barley.

8.2. Reliability of the Method

8.2.1. *Recovery.* Recovery trials were carried out with all-*rac*-α-tocopherol acetate in quantities of 0·17, 0·43 and 5·4 mg using full analysis (saponification, extraction as described above). The recovery rates for α-tocopherol from aliquots of the extract (analogue 7.3) were between 92·2 and 101·5%, giving an average of 97·3% ($N = 26$). Given the high recovery rates no recovery correction of the result (8.1) was generally necessary. If required it can be determined by the addition of all-*rac*-α-tocopherol acetate to the weighed portion (6.2.1) of the order of the basic content found.

8.2.2. *Repeatability.* The analysis of a coarse wheat meal enriched with 10 ppm all-*rac*-α-tocopherol acetate gave the following figures for 10 full analyses ($N = 10$) calculated as α-tocopherol: $\bar{x} = 13\cdot6$ mg α-tocopherol/kg; $s = 0\cdot9117$ mg/kg; $s_{rel} = 6\cdot7\%$.

8.2.3. *Comparability.* This was not determined since there are no syndicate tests.

8.2.4. *Comparison with other methods for determining α-tocopherol.* No comparisons are available.

9. ANALYSIS REPORT

The result of the determination is to be given with a reference to this method. Any operations which are not mentioned in the method should be listed.

10. REFERENCES

Remark: The method described above represents the current status (November 1983) of the method published by Manz and Philipp (Ref. 10.6), which has been considerably simplified here. The references cited provide information exclusively on the use of the HPLC method—with direct and reverse phase and UV—or fluorescence detection—for the determination of tocopherol in foodstuffs in the broadest sense. The far more extensive literature on the determination of tocopherol in oils and fats by means of HPLC is not cited here since this would go far beyond the borders of reference literature and since in principle, apart from the possible omission of saponification, there are no differences in tocopherol analysis in foodstuffs.

10.1. P. J. Barnes, *Z. Lebensm. Unters. Forsch.*, **174**, 467–71 (1982).
10.2. P. J. Barnes, *Fette-Seifen-Anstrichmittel*, **84** (7), 256–69 (1982).
10.3. E. Sanzini and G. Bellomonte, *Acta Vitaminol. Enzymolog.*, **4** (4), 347–52 (1982).
10.4. P. J. Barnes and P. W. Taylor, *Chemistry in New Zealand*, **45** (6), 176, 178, 180 (1981).
10.5. K. Hirai, S. Ishizawa, M. Miura, T. Hara and K. Owada, *Osakashiritsu Daigaku Seikatsukagakubu Kiyo*, **29**, 9–13 (1981).
10.6. U. Manz and K. Philipp, *Intern. J. Vit. Nutr. Res.*, **51** (4), 342–8 (1981).
10.7. U. Coors and A. Montag, *Chem. Mikrobiol. Techn. Lebensm.*, **7**, 21–2 (1981).
10.8. P. J. Barnes and P. W. Taylor, *J. Sci. Food Agric.*, **31**, 997–1006 (1980).
10.9. P. J. Barnes, *Food Flavourings, Ingredients and Processing*, **2** (4), 19 (1980).
10.10. S. S. O. Hung, Y. C. Cho and St. J. Slinger, *J. Assoc. Off. Anal. Chem.*, **63** (4), 889–93 (1980).
10.11. J. N. Thompson and G. Hatina, *J. Liq. Chromatogr.* **2** (3), 327–44 (1979).
10.12. W. A. Widicus and J. R. Kirk, *J. Assoc. Off. Anal. Chem.*, **62**, 637–41 (1979).
10.13. H. Fukuba, T. Miyoshi and T. Tsuda, *J. Am. Oil Chem. Soc.*, **56** (2), A 189 (1979).
10.14. P. Söderhjelm and B. Andersson, *J. Sci. Food Agric.*, **29**, 697–702 (1978).
10.15. P. J. van Niekerk and L. M. du Plessis, *South African Food Review*, 167–71 (June 1976).

9
Free Tocopherols and Tocotrienols (Vitamin E) in Edible Oils and Fats: HPLC Method*

1. PURPOSE AND SCOPE

The method describes a procedure for the qualitative and quantitative determination of the individual tocopherols. It is applicable to edible vegetable oils and fats. The method is not applicable to fatty food preparations such as the fatty phase of margarine or mayonnaises and blends of edible fats, since they might contain esterified tocopherols.

2. DEFINITION

Individual tocopherols determined by this method are taken to mean the four tocopherols (α-, β-, γ- and δ-T) and four tocotrienols (α-, β-, γ- and δ-T$_3$) (see Fig. 8). These vitamers occur mainly in vegetable oils and fat. Their content is given in mg/kg.

3. BRIEF DESCRIPTION (PRINCIPLE OF THE METHOD)

The sample is dissolved in acetone, refrigerated to $-80\,°C$ and filtered. The mixture of tocopherols contained in the filtrate is separated by HPLC adsorption chromatography; the tocopherols are detected by UV- or fluorimetric methods. They are identified by comparing their retention times with those of the appropriate pure substances. For their quantitative determination the peak areas are integrated and compared with those of external standards.

---

*DGF Einheitsmethode F-II 4/84.

4. CHEMICALS

4.1. L-Ascorbyl palmitate, e.g. Roche or Merck.
4.2. Acetone, puriss. (for acetone/dry ice bath).
4.3. Acetone, for chromatography, with addition of 0·1 mg of ascorbyl palmitate (4.1) per litre.
4.4. Dry ice (solid CO_2).
4.5. Nitrogen, oxygen-free, 99·99% (v/v).
4.6. 1,4-Dioxane, for chromatography, stabilised with approx. 1·5 mg/kg BHT (2,6-di-*tert*-butyl-4-methylphenol).

Note 1: 1,4-Dioxane is suspected of being carcinogenic. The present 'MAK' value is 50 ml/m^3.

4.7. *n*-Heptane or *n*-hexane, for chromatography, stabilised with approx. 1·5 mg/kg BHT.
4.8. *tert*-Butyl-methyl ether, for chromatography.
4.9. Mobile phases (either of the following).
  4.9.1. 1,4-Dioxane in *n*-hexane (*n*-heptane); adjust 40 ml dioxane (4.6) with *n*-hexane or *n*-heptane (4.7) to 1000 ml.
  4.9.2. *tert*-Butyl-methyl ether in *n*-hexane (*n*-heptane); adjust 30 ml of the ether (4.8) with *n*-hexane or *n*-heptane (4.7) to 1000 ml. The required amount of 1,4-dioxane or *tert*-butyl methyl ether respectively is placed in the 1000 ml volumetric flask (5.2) and topped up with *n*-hexane (*n*-heptane). Immediately before the mobile phase is used it is degassed in the ultrasonic bath (5.3) for about 3 min.

Note 2: Pay attention that an accurate concentration of 3% (v/v) of the ether (4.8) in the mobile phase (4.9.2) is used; increasing concentrations of the ether (4.8) may cause a change in the sequence of elution of the individual tocopherols from the column.

4.10. α-, β-, γ-, δ-Tocopherols as reference substances.

Note 3: Reference tocopherols are available from E. Merck (POB 4119, D-6100 Darmstadt 1) (Article No. 15496 'Tocopherol Isomere' (alpha, beta, gamma, delta für biochemische Zwecke). Before 1985 the content of each vitamer in this set of tocopherols was not warranted and had therefore to be determined by the present HPLC method before use. It may also differ from set to set. The purity was in general of the order of 90–100%. However, purity data based on gas chromatography are not relevant for the present method. The purity has now been improved.

5. APPARATUS AND AUXILIARY EQUIPMENT

5.1. Analytical balance.
5.2. Volumetric flask, 1000 ml.
5.3. Ultrasonic bath.
5.4. Volumetric flasks (amber glass), 10, 25, 50, 100, 200, 1000 ml.
5.5. Erlenmeyer flask, 100 ml.
5.6. Dewar vessel, approx. 2·5 litre.
5.7. Low-temperature suction filter (see Fig. 15).
5.8. Rotary evaporator with water bath.
5.9. Test tubes, approx. 30 ml.
5.10. Two round-bottomed flasks, standard ground joint 29/32, 250 ml.
5.11. HPLC apparatus comprising:
    5.11.1. Basic HPLC equipment, consisting of injection system, high pressure pump(s) and analytical column (5.11.5).
    5.11.2. HPLC detector, either:
        5.11.2.1. Spectrophotometer: with variable wavelength, set to 295 nm.
        5.11.2.2. Spectrofluorimeter: excitation wavelength set to 295 nm, emission to 330 nm.
    5.11.3. Recorder, with compensator.
    5.11.4. Integrator.
    5.11.5. Analytical HPLC column; stationary phase Si 60, 5 µm, 250 × 4 mm; for straight phase chromatography.
5.12. Microlitre syringe, 100 µl; graduated in 1 µl divisions.
5.13. Transfer pipettes, 10 ml.
5.14. Ground-glass tapered flask, standard ground joint 14, 5, 10 ml.
5.15. Pipette with a long capillary tip, 1 ml.

6. SAMPLE

6.1. *Sampling*

Samples are taken in accordance to the DGF Einheitsmethoden C-I 1 to C-I 5.

6.2. *Sampling Method and Preparation of the Sample*

Samples are used directly without any further preparation or processing.

    6.2.1. *Weighed portion.* Approx. 1 g of oil or fat.

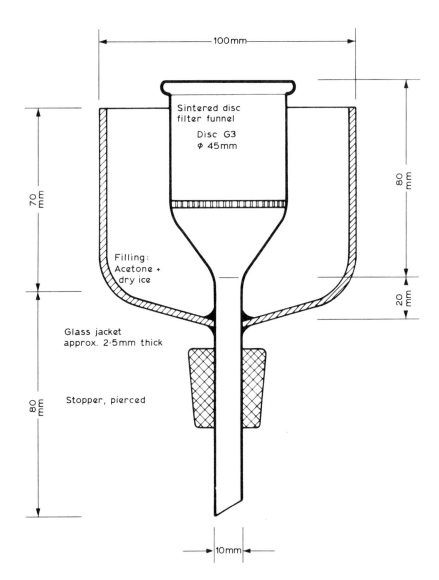

Fig. 15. Deep freezing suction filter.

7. PROCEDURE

7.1. Pre-purification of Individual Tocopherols

7.1.1. Approx. 1 g of the sample, accurately weighed to 0·001 g, is transferred to a 100 ml Erlenmeyer flask (5.5) and dissolved in approx. 50 ml acetone (4.3).

7.1.2. The solution is cooled to approx. $-80\,°C$ for approx. 10 min with acetone (4.2) and dry ice (4.4) in the Dewar flask.

7.1.3. At the same time four test tubes (5.9), containing 25 ml acetone (4.3) each, are subjected to intensive cooling ($-80\,°C$).

7.1.4. The frozen content of the Erlenmeyer flask is shaken thoroughly and, with the help of the filter apparatus (5.7), sucked through the G3 filter disc, precooled to $-80\,°C$, into a 250 ml round-bottomed flask (5.10). The crystalline residue is washed with two portions, each of 25 ml, of the precooled acetone (7.1.3).

7.1.5. The filled 250 ml round-bottomed flask is replaced by an empty 250 ml round-bottomed flask.

7.1.6. The frozen residue is redissolved in approx. 50 ml acetone (4.3) warmed up to about 30 °C and sucked into the empty round-bottomed flask, the solvent being added in small portions.

7.1.7. With the solution (7.1.6) the procedure of freezing, suction and rewashing is repeated once more as described in sections 7.1.2 to 7.1.4.

7.1.8. The second filtrate is combined with the first in the 250 ml round-bottomed flask (7.1.4); the empty receiver is rinsed twice with 25 ml acetone (4.3) and the rinsing fluid is added to the combined filtrate.

7.1.9. The combined acetone extracts are evaporated to dryness by applying a weak vacuum in a rotary evaporator (5.8) flushed with nitrogen (4.5). At completion the vacuum is replaced with nitrogen.

7.1.10. The dry residue is quantitatively transferred using approx. 5 ml n-hexane (4.7) into a 10 ml volumetric flask (5.4) and adjusted with the same solvent to the mark (= sample test solution). If in the course of this operation the solution becomes turbid, this has to be filtered into the graduated flask. This filtration must be indicated in the analysis report (section 9).

7.2. Separation of the Individual Tocopherols

7.2.1. Depending on the presumed concentration of tocopherols in the sample 10 to 100 µl of the sample test solution (7.1.10) are injected onto the analytical column of the HPLC apparatus (5.11).

7.2.2. The tocopherols are eluted with one of the mobile phases indicated (4.9.1 or 4.9.2).

7.2.3. The volume of the injection fluid and the flow rate of the mobile phase have to be adjusted to the actual HPLC line; they have to be varied until optimal conditions are obtained for separation of all tocopherols and tocotrienols.

7.2.4. The sequence in which the separate tocopherols should appear in the eluate is the following: α-T, α-T$_3$, β-T, γ-T, β-T$_3$, γ-T$_3$, δ-T and δ-T$_3$.

Note 4: The chromatogram represented in Fig. 16 may serve as a guideline.

7.3. Plotting of the Calibration Curves

7.3.1. For dissolving the reference substances (4.10) a dried 10 ml ground-glass tapered flask (5.14) is weighed accurately to within 0·2 mg.

7.3.2. Ampoules containing approx. 50 mg of the corresponding reference tocopherols are opened with the help of an ampoule saw. Approx. 1 ml *n*-hexane (4.7) is injected into the opened ampoule and after approx. 30 s the solution is transferred into the weighed tapered flask (7.3.1). This operation is repeated four times with approx. 1 ml *n*-hexane each time.

7.3.3. The solvent is removed from the solution (7.3.2) at 40 °C and approx. 30 torr (water jet pump) in the rotary evaporator (5.8). After the removal of the solvent the evaporator is flushed with nitrogen (4.5), the tapered flask is disconnected, cooled to room temperature and weighed accurately to within 0·2 mg.

7.3.4. The residue of method 7.3.3 is dissolved in *n*-hexane (4.7) and quantitatively transferred to a 50 ml volumetric flask (5.4) and adjusted with the same solvent.

7.3.5. Ten millilitres of each of the solutions (7.3.4), containing the tocopherols in question, are pipetted into 25, 50, 100 and 200 ml volumetric flasks (5.4) and adjusted with *n*-hexane.

7.3.6. When the calibration curves are being plotted the purity of the reference tocopherols must be taken into consideration (see section 7.2.4, note).

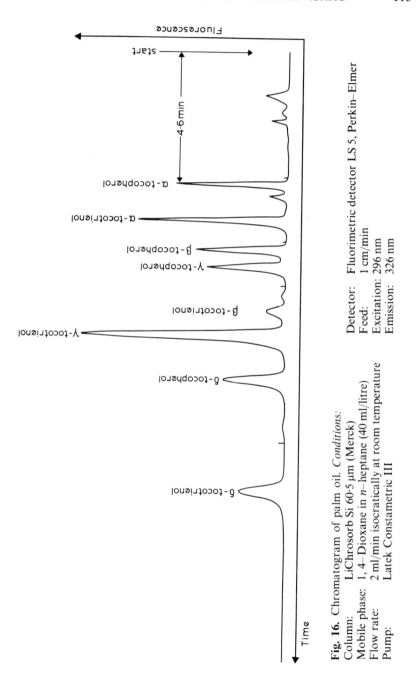

Fig. 16. Chromatogram of palm oil. *Conditions:*
Column: LiChrosorb Si 60·5 μm (Merck)
Mobile phase: 1,4–Dioxane in *n*–heptane (40 ml/litre)
Flow rate: 2 ml/min isocratically at room temperature
Pump: Latek Constametric III

Detector: Fluorimetric detector LS 5, Perkin–Elmer
Feed: 1 cm/min
Excitation: 296 nm
Emission: 326 nm

7.3.7. Equal volumes of each of the solutions (7.3.5) are injected into the chromatography column and eluted with the appropriate flow rate (7.2.3) and the mobile phase selected.

7.3.8. The integrated peak areas are plotted graphically as a function of the injected mass of the corresponding reference tocopherols; the points are joined to produce calibration curves.

7.3.9. The four calibration curves for α-, β-, γ- and δ-T are assumed to be equally valid for the corresponding tocotrienols.

Note 5: The calibration curves plotted according to section 7.3 are only valid for the HPLC system used. They depend to a high degree on the individual differences in functional details of a certain assembly of an HPLC line. Therefore it is not possible to indicate generally valid calibration curves or response factors for the different tocopherols and tocotrienols. However, the following approximate response factors, which were estimated from some syndicate tests (fluorimetry, *n*-hexane or *n*-heptane as solvents), may serve as guidelines: α-tocopherol, 1·00; β-tocopherol, 0·8; γ-tocopherol, 0·8; δ-tocopherol, 0·7.

Plotting of the calibration curves (7.3) has to be repeated from time to time, especially just before quantitative determinations of individual tocopherols.

8. EVALUATION

8.1. *Qualitative Evaluation*

8.1.1. Peaks of the chromatograms resulting from method 7.2 are identified as T-peaks by comparing their retention times with those of the α-, β-, γ- and δ-reference tocopherols from method 7.3.

8.1.2. In case of doubt, reference tocopherols are added to the prepurified sample (7.1) and the mixture is chromatographed again according to method 7.2.3.

8.1.3. After identification of the tocopherols by means of their retention times the presence of tocotrienols is deduced from the sequence of the other peaks appearing in the eluate (7.2.4).

8.2. *Quantitative Evaluation*

The concentration for each of the separate tocopherols is calculated according to the equation:

$$c = \frac{m_1 v_1}{m_0 v_2}$$

where: c = concentration of the corresponding tocopherol (mg/kg); m_0 = weighed portion of the sample (g); m_1 = amount of the corresponding tocopherol, read from the calibration curves (µg); v_1 = volume of the sample test solution (7.1.10) (µl); v_2 = injected volume of the sample test solution in (7.2.1) (µl).

8.3. Reliability of the Method

8.3.1. *Repeatability*. The difference between two determinations made one after the other on the same day by the same analyst with the same equipment and the same sample must not exceed 26% (relative) at a concentration of individual tocopherols of less than 200 mg/kg and not exceed 12% at a concentration of more than 200 mg/kg.

8.3.2. *Comparability*. The difference between two determinations made on the same sample by two different laboratories must not exceed 43% (relative) at a content of individual tocopherols of less than 200 mg/kg and not exceed 24% at a content of more than 200 mg/kg.

9. ANALYSIS REPORT

The result of the determination is to be given with a reference to this method, to the used mobile phase and the other conditions of chromatography. A filtration carried out at point 7.1.10 must be indicated. All data referring to an identification of the sample and any operations not mentioned in the method must be entered in the analysis report.

10. REFERENCES

The described method has been published as DGF 'Einheitsmethode' F-II 4/84; see Ref. 10.1. It is the result of seven extensive syndicate tests conducted by the DGF between 1979 and 1983. The manuscript was released and made available to COST 91 by the Deutsche Gesellschaft für Fettwissenschaft (DGF), Soester Str. 13, D-4400 Münster/Westfalen, Bundesrepublik Deutschland.

10.1. M. Arens, S. Kroll and W. Müller-Mulot, *Fette-Seifen-Anstrichmittel*, **84** (4), 148–51 (1984).

PART III

TENTATIVE METHODS

10
Vitamin B$_2$ (Riboflavin) in Foodstuffs: HPLC Method

1. PURPOSE AND SCOPE

The method describes a procedure for the quantitative determination of vitamin B$_2$ (riboflavin) in foodstuffs, that is the naturally occurring vitamin B$_2$ and, if required, the vitamin B$_2$ added during the manufacture of the foodstuff. The method is applicable to fresh and stored foodstuffs intended for immediate consumption such as grain, flour, pastry and confectionery, fresh milk and dried milk products, vegetables and complete meals.

Vitamin B$_2$ contents above 30 µg per 100 g of sample can be determined quantitatively, and contents of 10 µg per 100 g can be estimated semi-quantitatively.

2. DEFINITION

Vitamin B$_2$ content is understood to be the vitamin B$_2$ content, calculated as riboflavin, determined by the procedure described here. It is given in µg of riboflavin per 100 g of sample.

3. BRIEF DESCRIPTION (PRINCIPLE OF THE METHOD)

After acidic disintegration and enzymatic digestion which releases the riboflavin, the sulphuric acid extract is diluted with methanol and water; the resulting riboflavin concentrate is analysed directly by means of HPLC reverse phase on an RP 18 column—the peak areas are evaluated against an external riboflavin standard which has undergone the same analytical procedure as the sample.

4. CHEMICALS

Remark: Unless otherwise specified, AR grade chemicals are to be used; water must be distilled and taken from glass vessels or be of corresponding purity.

4.1. Sodium acetate trihydrate, puriss. >99·5%, e.g. Fluka or Merck.
4.2. Clara-Diastase (Clarase 300), e.g. Fluka, Article No. 27540.
4.3. Riboflavin for biochemical purposes, 99% of the dry substance, e.g. Merck; dried over P_2O_5 and stored.
4.4. Diammonium hydrogen phosphate (secondary ammonium phosphate), e.g. Merck.
4.5. Sulphuric acid, 95–97%, e.g. Merck.
4.6. Sulphuric acid, approx. 2·5 M (5·0 N) solution; dilute 137·5 ml H_2SO_4 (4.5) with water to 1000 ml. Where required, 0·4 and 0·2 M (0·8 and 0·4 N) solutions are made up by dilution.
4.7. Sulphuric acid approx. 0·1 M (0·2 N) solution; dilute 5·5 ml H_2SO_4 (4.5) with water to 1000 ml.
4.8. Sulphuric acid approx. 0·01 M (0·02 N) solution; dilute 1 volume of 0·1 M H_2SO_4 (4.7) with 9 volumes of water.
4.9. Methanol for chromatography, Lichrosolv, e.g. Merck.
4.10. 1,4-Dioxane, puriss., stabilised with BHT, e.g. Merck.
4.11. Petroleum ether, puriss., boiling range 50–70 °C, e.g. Merck.
4.12. Solutions.
    4.12.1. Clara-Diastase suspension, 10%; 5 g of Clara-Diastase (4.2) is made up to 50 ml with water and mixed with a magnetic stirrer; it is made up fresh immediately before use.
    4.12.2. Sodium acetate solution, approximately 2·5 M; 340 g of sodium acetate (4.1) are dissolved in water to 1000 ml.
    4.12.3. Diammonium hydrogen phosphate solution; dissolve 2·5 g of the phosphate (4.4) in 650 ml water (constituent of the mobile phase).
    4.12.4. *Vitamin B_2 standard solution.* Twenty-five milligrams of riboflavin (4.3) are dissolved in 0·01 M (0·02 N) sulphuric acid (4.8) in a 500 ml graduated flask and made up to the mark (=stock solution with 50 µg riboflavin/ml). Two millilitres of stock solution (=100 µg riboflavin) are made up to the mark with 0·01 M (0·02 N) sulphuric acid. (4.8) in a 500 ml graduated flask (=standard solution with 0·2 µg riboflavin/ml=200 ng/ml).

Note 1: The stock solution can be stored without losses for up to three months under toluene at +4 °C in the dark. The standard solution can be kept for four weeks at +4 °C.

5. APPARATUS AND AUXILIARY EQUIPMENT

5.1. Standard basic laboratory equipment.
5.2. Small autoclave or pressure cooker.
5.3. Water bath with thermostat.
5.4. Polytron rod or similar mixer; scissors, knife.
5.5. Mixer, e.g. Waring blender.
5.6. Refrigerator.
5.7. Pleated filter, e.g. Schleicher and Schüll No. 589^3 1/2.

Note 2: Only filters which adsorb no vitamin B_2 can be used. If in doubt a check should be run.

5.8. Laboratory centrifuge.
5.9. HPLC apparatus, e.g. Perkin–Elmer, Waters, etc.
Detector, e.g. Perkin–Elmer 650–10 Fluorimeter; 453/521 nm.
Integrator, e.g. Spectra Physics 4100.
Column, e.g. LiChrosorb RP-18, 5 µm, 125 × 4 mm, Merck (Hibar) or µ-Bondapak C_{18}, 10 µm, 300 × 3·9 mm, Waters.
5.10. Pre-column with Perisorb RP-18 (30–40 µm), Merck (to be used if necessary).
5.11. Filtering apparatus for the mobile phase, e.g. Millipore 0·45 µm.

6. SAMPLE

6.1. *Sampling*

The proportions and composition of the sample to be taken must be representative of the material to be analysed. The sample must, in any case, be taken from a sizeable initial amount.

6.2. *Sampling Method and Preparation of the Sample*

In many cases (with perishable, fresh foodstuffs, in field tests, etc.) samples must be *preserved before analysis*. In such cases, the sample, which may be deep-frozen for short term preservation and then carefully defrosted at +4 °C, should be broken down with 0·4 M, 0·2 M or 0·1 M sulphuric acid (4.6–4.7) and homogenised for between 20 s and 2 min with the Polytron rod (5.4). The ratio between the sample and sulphuric acid should be selected so that the homogenate flows freely after homogenisation and the final acid concentration is approximately 0·1 M.

The following are examples of quantities to be employed. For 90–100 g of soup, juice, milk and other samples made up largely of water, 0·4 M sulphuric acid; for 60–70 g meat, vegetables, fruit, etc., 0·2 M sulphuric acid; for 20–30 g bread, flour, pastry and other strongly swelling products with low moisture content, 0·1 M sulphuric acid. In the case of dried products swelling strongly in water such as crisp bread, a greater degree of dilution must be provided for. *Preserved samples* can be kept at −20 °C for up to three months.

Samples which are not to be preserved, such as dry non-perishable foodstuffs, are immediately taken for disintegration in 0·1 M sulphuric acid (4.7) after having been comminuted and homogeneously mixed (5.5).

6.2.1. *Weighed portion.* The weighed portion will depend on the amount of vitamin B_2 expected; in general it will be 10 g of sample (in the original state or in the form of a sulphuric acid homogenate) and will then provide between 3 and 25 µg vitamin B_2. Further dilution is required with higher vitamin B_2 contents.

7. PROCEDURE

7.1. *Remark*
Except where indicated, all the analytical steps of the method must be performed without interruption. Duplicate determinations must be carried out on the same day.

7.2. *Disintegration and Digestion*

7.2.1. *Disintegration with sulphuric acid.* Approximately 10 g of the sample in the original condition or homogenised in 0·1 M sulphuric acid (4.7), weighed accurately to 0·01 g, are made up to 40 g with the same sulphuric acid (4.7) in a 250 ml wide-neck conical flask and treated for 15 min at 120 °C in the autoclave (5.2). After cooling to room temperature, 6–7 ml sodium acetate solution (4.12.2) are added in order to adjust the pH value to $4·5 \pm 0·1$.

7.2.2. *Enzymatic digestion* (*obtaining the extract*). A volume of 5 ml of the 10% Clara-Diastase suspension (4.12.1) is added to the solution (7.2.1), and the conical flask allowed to incubate for 60–90 min at 45 °C in the water bath (5.3). After cooling to room temperature, the digested mixture is acidified with 4 ml of 2·5 M sulphuric acid (4.6), and the contents of the flask are transferred to a 100 ml graduated flask, after rinsing with water, and made up to the mark. After filtering through a pleated filter (5.7), the initial filtrate having been rejected, the filtered extract is obtained.

> *Note 3:* With regard to bound riboflavin and incubation time in the case of enzymatic digestion, see also Ref. 10.4. It is advisable to check whether an increased incubation time (2 h at 45 °C or overnight at 37 °C) will increase the riboflavin yield still further. The clear filtered extract can be stored overnight at 4 °C in the refrigerator (5.6) for further processing.

7.3. Sample Test Solution

The extract (7.2.2) or an aliquot of the extract is diluted with a solution of methanol (4.9) and water (7:3, v/v) in a ratio of 1:1 (v/v) and centrifuged for 15 min at 6000 rpm (5.8) (=sample test solution with 15–125 ng riboflavin/ml, which corresponds to 0·3–2·50 ng/20 µl). An ultrafilter (5.11) can be used for filtering instead of centrifuging. The extract must be clear.

7.4. Standard Test Solutions

7.4.1. Standard test solution I. A volume of 4·0 ml vitamin B_2 stock solution (4.12.4) is diluted with 0·1 M sulphuric acid (4.7) to 100 ml. Ten millilitres of solution (corresponding to 20 µg riboflavin) are subjected to the complete analysis procedure (7.2.1 and 7.2.2) in a matrix-free condition. Result: 100 ml 'extract' with 0·2 µg riboflavin/ml (=200 ng/ml). The 'extract' is diluted 1:1 (v/v) as described in section 7.3 (=standard test solution I with, in theory, 100 ng/ml, corresponding to 2 ng/20 µl).

> *Note 4:* If the weighed portion contains less than 10 µg riboflavin, 10 µg riboflavin (instead of 20 µg) should be subjected to the full analysis for standard test solution I. A standard test solution I will then result with 50 µg riboflavin/ml, corresponding to 1 ng/20 µl.

7.4.2. Standard test solution II. A volume of 10·0 ml vitamin B_2 stock solution (4.12.4), corresponding to 2 µg riboflavin, is diluted in the same way as for solution I (7.4.1), but without the analysis procedure, with methanol (4.9)/water (7:3, v/v) in the ratio 1:1 (v/v) (=standard test solution II with 2 ng riboflavin/20 µl).

> *Note 5:* Standard test solution II can be adjusted to 1 ng riboflavin/20 µl, with the vitamin B_2 standard solution (4.12.4) being diluted with 0·01 M sulphuric acid (4.8) (1:1, v/v) (giving 100 ng/ml) and this solution then

being diluted with methanol (4.9)/water (7:3) in the ratio 1:1 (v/v), so that the standard test solution II then contains 1 ng/20 µl.

The two standard test solutions I and II (with and without preliminary analysis) should be used together for all series of measurements; the results for solution I should be at the most 30% below those for solution II. If they are further apart the whole system should be checked. Solution II is intended exclusively for the calibration of the HPLC equipment and the stability control of the fluorimeter. The evaluation is done with solution I.

7.5. HPLC

7.5.1. Twenty microlitres of the sample test solution (7.3) and standard test solutions I (7.4.1) and II (7.4.2) are subjected to isocratic chromatography with reverse phase HPLC. Calibration and stability control of the fluorimeter are checked at the beginning with standard test solution II (7.4.2) and during the course of the analysis (every sixth injection).

7.5.2. *Chromatographic conditions:*
Apparatus (5.9).
Stationary phase: LiChrosorb RP-18, 5 µm or µ-Bondapak C_{18}, 10 µm.
If necessary pre-column (5.10).
Injection volume: 20 µl.
Mobile phase (eluant): 350 ml methanol (4.9) with 650 ml $(NH_4)_2HPO_4$ solution (4.12.3) and 10 ml dioxane (4.10) are mixed and ultrafiltered (5.11).
Pressure: between 110 and 160 bars.
Flow-rate: 1·0 ml/min.
Detection: fluorimetry: excitation wavelength, 453 nm; emission wavelength, 521 nm; retention times for riboflavin were, according to type of column, 3·5–3·8, 4·3–4·4 and 6·1–6·4 min.
Peak evaluation: by integrated areas.

Note 6: The use of a fluorescence detector enables vitamin B_2 to be determined in a more specific fashion; extraneous fluorescences are separated by the previous chromatography. The composition of the mobile phases is selected so that vitamins B_1 and B_2 are properly separated from each other.

7.5.3. *Chromatograms.* Figures 17 and 18 show HPLC chromatograms for riboflavin determinations in a standard solution.

8. EVALUATION

8.1. Calculation

The riboflavin content in the sample test solution (7.3) is calculated using the external standard. The initial calibration of the HPLC line is done with 20 µl of the standard test solution II (7.4.2) which corresponds to 2 ng of riboflavin. At the beginning of a measurement series, three measurements with standard test solution I (7.4.1) originating from a single batch are carried out; the peak areas (average from three measurements) are assigned to the theoretical content of 2 ng riboflavin/20 µl. The peak areas of the sample test solutions are compared with that of standard I. Account must be taken of the weighed portion, the removal of aliquots and dilution.

8.1.1. Range of linearity.
The linear proportionality of the peak area values for the sample test solutions in comparison with those of the standard are in the range 0·1–4·0 ng riboflavin/20 µl, corresponding to 5–200 ng/ml sample test solution.

8.1.2. Detection limit.
Thirty micrograms of riboflavin in 100 g sample or 15 ng/ml sample test solution can be quantitatively evaluated; at low concentration, semi-quantitative determination is possible. The detection limit of the HPLC method in comparison with conventional methods for vitamin B_2 is some ten times better, while it is approximately twice as good as microbiological methods (*Lactobacillus casei*).

8.2. Reliability of the Method

8.2.1. Recovery.
Separate recovery determinations are superfluous since the use of the standard test solution I (7.4.1) corrects for all losses during the procedure. The actual recovery rate is estimated to be between 70 and 80%.

8.2.2. Repeatability.
The following standard deviations were found for three measurement samples (white and brown flour) with riboflavin contents of between 100 and 125 µg/100 g: standard deviation: 3–8 µg/100 g; $s_{rel}\%$, 2·4–6·4%.

8.2.3. Comparability.
This was not determined since there are no syndicate tests.

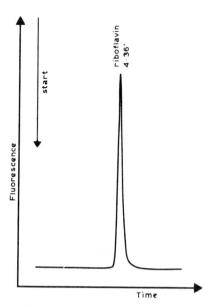

Fig. 17. Chromatogram of a standard riboflavin solution (0·206 µg/ml, injection: 20 µl corresponding to 4·12 ng).

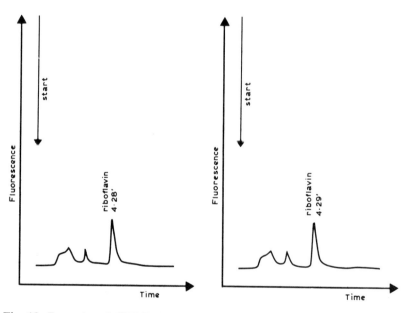

Fig. 18. Examples of HPLC chromatograms of two wheat flour extracts (injection: 20 µl sample test solution).

8.2.4. *Comparison with other methods for determining vitamin B_2.* No comparable measurements on identical types of samples are available. An HPLC Method (Ref. 10.3) put forward by Professor G. Testolin, of the University of Milan, should be mentioned in this connection: following sulphuric acid and enzymatic (diastase–papain) hydrolysis, the water-diluted hydrolysate is separated directly on an ODS Ultrasil column with a methanol/phosphate buffer, pH 5, eluant (3·5:6·5); fluorimetry is carried out at 390/475 nm. The external standard contains 4 ng riboflavin/20 µl injection volume.

9. ANALYSIS REPORT

The result of the determination is to be given with a reference to this method. Operations not mentioned in the described method must be indicated.

10. REFERENCES

10.1. Dr W. Schüep, Report, F. Hoffmann-La Rock & Co. AG, Basel, January 1982.
10.2. Dr P. Scheffeldt and H. P. Meier, Laboratory method, Institut für Ernährungsforschung, Seestrasse 72, CH-8803 Rüschlikon-Zürich, July 1983.
10.3. Professor G. Testolin, Laboratory method, Fisiologia Nutrizione, Università di Milano, Via Celoria 2, Milano, June 1983. Private communication.
10.4. R. Rettenmaier and J.-P. Vuilleumier, *Intern. J. Vit. Nutr. Res.*, **53** (1), 32–5 (1983).
10.5. S. H. Ashoor, G. J. Seperich, W. C. Monta and J. Welty, *J. Food Sci.*, **48** (1), 92–4, 110 (1983).
10.6. E. A. Woodcock, J. J. Warthesen and T. P. Labuza, *J. Food Sci.*, **47**, 545–9, 555 (1982).
10.7. J. K. Fellman, W. E. Artz, P. D. Tassinari, C. L. Cole and J. Augustin, *J. Food Sci.*, **47** (6), 2048–50, 2067 (1982).
10.8. A. Bognár, *Deutsche Lebensmittel-Rundschau*, **77** (12), 431–6 (1981).
10.9. I. D. Lumley and R. A. Wiggins, *Analyst (London)*, **106** (1267), 1103–1108 (1981).
10.10. J. F. Kamman, T. P. Labuza and J. J. Warthesen, *J. Food Sci.*, **45** (6), 1497–9 (1980).
10.11. P. J. Richardson, D. J. Favell, G. C. Gidley and A. D. Jones, *Proc. Anal. Div. Chem. Soc.*, **15**, 53–5 (1978).

10.12. D. R. Osborne and P. Voogt, *The Analysis of Nutrients in Foods*, Academic Press, London, 1978, p. 82.
10.13. A. T. Rhys Williams and W. Slavin, *Chromatog. Newsletter*, **5** (1), 9–11 (1977).
10.14. P. Nielsen, P. Rauschenbach and A. Bacher, *Anal. Biochem.*, **130** (2), 359–68 (1983).

11
Vitamin B₆ in Foodstuffs: HPLC Method

1. PURPOSE AND SCOPE

The method describes a procedure for the quantitative determination of vitamin B_6—pyridoxamine (PAM), pyridoxal (PAL) and pyridoxine (POL)—in foodstuffs. It is used to determine the naturally occurring vitamin B_6 vitamers—pyridoxamine, pyridoxal and pyridoxine (pyridoxol)—and any vitamin B_6 (pyridoxine) added in the course of food manufacture.

The method is applicable to both fresh and stored foodstuffs intended for immediate consumption, such as cereals, flour, pasta products, bakers' wares, baby food, vegetables, meat and meat products, milk and milk products as well as to individual dishes and complete meals.

Pyridoxamine, pyridoxal and pyridoxine contents of not less than 10 µg/100 g of sample can be determined quantitatively.

2. DEFINITION

Vitamin B_6 content is taken to mean the content of pyridoxamine, pyridoxal and pyridoxine determined in accordance with the method described below. It is given in mg of pyridoxine (pyridoxol)/100 g of sample

1 mg pyridoxine = 1·008 mg pyridoxamine = 1·012 mg pyridoxal.

3. BRIEF DESCRIPTION (PRINCIPLE OF THE METHOD)

After acidic disintegration of the sample material with the release of vitamin B_6 (pyridoxamine, pyridoxal and pyridoxine), the sulphuric acid extract is diluted with water.

Pyridoxamine, pyridoxal and pyridoxine are separated directly by reversed-phase HPLC on a Spherisorb ODS column and measured fluorimetrically.

Note 1: Separation of pyridoxamine (PAM), pyridoxal (PAL) and pyridoxine (POL) is necessary because of the different fluorescent activities of the individual vitamin B_6 vitamers. For example, the same quantity of pyridoxamine exhibits 2·8 and 3 times as much fluorescence as pyridoxal and pyridoxine respectively.

4. CHEMICALS

Remark: Unless otherwise specified, AR grade chemicals are to be used. Water must be either distilled from quartz-glass containers or be of equivalent purity.

4.1. Sulphuric acid, 95–97%, e.g. Merck.
4.2. Sulphuric acid, approx. 0·5 M (approx. 1 N) solution; dilute 27·5 ml H_2SO_4 (4.1) with water to 1000 ml.
4.3. Sulphuric acid, approx. 0·1 M (approx. 0·2 N) solution; dilute 200 ml 0·5 M H_2SO_4 (4.2) with water to 1000 ml.
4.4. Mobile phase (eluant): sulphuric acid, approx. 0·04 M (approx. 0·08 N) solution; dilute 400 ml 0·1 M H_2SO_4 (4.3) with water to 1000 ml.
4.5. Methanol for use in chromatography, e.g. Merck.
4.6. Pyridoxamine dihydrochloride for biochemical purposes, e.g. Merck, Serva.
  4.6.1. Pyridoxamine stock solution (1 mg pyridoxamine/ml): dissolve 154·1 mg of pyridoxamine dihydrochloride (4.6) in 0·1 M sulphuric acid (4.3) with stirring and make up to 100 ml (stable for approximately four weeks in the refrigerator).
  4.6.2. Pyridoxamine standard solution I (100 µg pyridoxamine/ml); bring 10 ml of stock solution (4.6.1) in a 100 ml amber-glass volumetric flask (5.9) up to the mark with 0·1 M sulphuric acid (4.3); prepare fresh every week.
  4.6.3. Pyridoxamine standard solution II (10 µg pyridoxamine/ml): bring 10 ml of standard solution I (4.6.2) in a 100 ml amber-glass volumetric flask (5.9) up to the mark with 0·1 M sulphuric acid (4.3).
4.7. Pyridoxal hydrochloride for biochemical purposes, e.g. Merck.
  4.7.1. Pyridoxal stock solution (1 mg pyridoxal/ml): dissolve 121·8 mg

of pyridoxal hydrochloride (4.7) in 0·1 M sulphuric acid (4.3) with stirring and make up to 100 ml (stable for approximately four weeks in the refrigerator).

4.7.2. Pyridoxal standard solution I (100 µg pyridoxal/ml): bring 10 ml of stock solution (4.7.1) in a 100 ml amber-glass volumetric flask (5.9) up to the mark with 0·1 M sulphuric acid (4.3). Prepare fresh every week.

4.7.3. Pyridoxal standard solution II (10 µg pyridoxal/ml): bring 10 ml of standard solution I (4.7.2) up to the mark with 0·1 M sulphuric acid (4.3); the solution has to be prepared freshly before use.

4.8. Pyridoxine hydrochloride for biochemical purposes, e.g. Merck.

4.8.1. Pyridoxine stock solution (1 mg pyridoxine/ml): dissolve 121·6 mg of pyridoxine hydrochloride (4.8) in 0·1 M sulphuric acid (4.3) with stirring and make up to 100 ml (stable for approximately four weeks in the refrigerator).

4.8.2. Pyridoxine standard solution I (100 µg pyridoxine/ml): bring 10 ml of stock solution (4.8.1) in a 100 ml amber-glass volumetric flask (5.9) up to the mark with 0·1 M sulphuric acid (4.3); prepare fresh every week.

4.8.3. Pyridoxine standard solution II (10 µg pyridoxine/ml): bring 10 ml of standard solution I (4.8.2) in a 100 ml amber-glass volumetric flask (5.9) up to the mark with 0·1 M sulphuric acid (4.3); freshly prepare before use.

4.9. Paraffin oil, e.g. Merck.
4.10. Column material: Spherisorb RB ODS, 5 µm.
4.11. Petroleum ether, boiling range 40–60 °C.

5. APPARATUS AND AUXILIARY EQUIPMENT

5.1. Usual standard laboratory basic equipment.
5.2. Mixing apparatus, e.g. Waring blender, meat cutter, mincer.
5.3. Ultra Turrax or equivalent mixing device.
5.4. Laboratory autoclave or pressure cooker.
5.5. Magnetic stirrer.
5.6. pH meter.
5.7. Laboratory centrifuge.
5.8. Centrifuge tubes, 50 ml.
5.9. Amber-glass and other volumetric flasks, 100 ml.

5.10. Wide-necked polyethylene flasks with screw cap, 250 ml.
5.11. Folded filter, 18·5 cm in diameter, e.g. Macherey–Nagel 615 1/4.
5.12. Ultrafilter with spherical supporting container, e.g. Amicon Centriflo-Membrankegel, type CF 50.
5.13. HPLC apparatus, e.g. Kontron, Perkin–Elmer.
5.14. Integrator, e.g. HP, Perkin–Elmer.
5.15. HPLC spectrofluorimeter, e.g. Kontron, Perkin–Elmer.
5.16. Unpressurised sample admission valve with 50 µl loop, e.g. Rheodyne 7125.
5.17. Microlitre syringe, 100 µl, e.g. Hamilton.
5.18. Refined-steel HPLC column, e.g. Knauer.
Precolumn: 10×0.4 cm.
Main column: 25×0.4 cm.
5.19. Column filling device, e.g. Knauer.
5.20. HPLC columns: fill the precolumns and main columns (5.18) with Spherisorb RB ODS, 5 µm, in accordance with the instructions for the column filling device. Before use, condition the HPLC apparatus columns for 15 min with methanol (4.5) and for 15 min with the eluant (4.4) (flow rate 2 ml/min).

6. SAMPLE

6.1. *Sampling*
The proportions and composition of the sample to be taken must be representative of the material to be investigated (foodstuffs). A fairly substantial starting quantity must be taken in all cases (approximately five to ten times the amount of the weighed sample in the case of foodstuffs of heterogeneous composition).

6.2. *Sampling Method and Preparation of the Sample*
If necessary, the sample material must be comminuted and homogenised in the mixer, meat cutter or mincer (5.2). In many cases, (perishable, freshly-picked foodstuffs, fermentation tests, storage tests, etc.) samples must be preserved before analysis. For this purpose, an aliquot of the homogenised sample must be weighed in a polyethylene flask (5.10), mixed with 0·5 M sulphuric acid (4.2) and water, and homogenised with the Ultra Turrax (5.3) for 30 s. The homogeniser is then rinsed with water and the homogenate made up to 150 g (exact weight). The relative proportions of sample and sulphuric acid or water must be selected to

ensure that the homogenate flows freely after homogenisation and that the final acid concentration is approx. 0·1 M. The following are examples of quantities to be used. For 90–100 g of concentrated soup, juices, milk and other samples containing water, at least 30 ml 0·5 M sulphuric acid (4.2) and 20–30 ml of water; for 30–50 g of meat, meat products, vegetables, fruit, potatoes, refined pasta products, rice, etc., 30 ml 0·5 M sulphuric acid and 50–70 ml of water; for 30–40 g of bread, flour, pastry and other strongly swelling products with low moisture content, 30 ml 0·5 M sulphuric acid and 80–90 ml of water. Preserved products can be kept for three days in the refrigerator at 4 °C and for three months in the deep-freeze cabinet at −20 °C.

Samples which are not to be preserved, for example dry, non-perishable foodstuffs (baby food, etc.) are immediately diluted and mixed with 0·1 M sulphuric acid after having been comminuted and homogenised (5.3). It is advisable for samples with a high fat content to have the fat removed before being mixed with sulphuric acid (repeated digestion with petroleum ether (4.11); determination of the fat quantity extracted).

6.2.1. *Weighed portion.* The quantity of weighed portion will depend on the anticipated vitamin B_6 content. The portion is of the order of 5–10 g in the case of the untreated homogenate and of 5–40 g in the case of the sulphuric acid homogenate (6.2). The pyridoxamine, pyridoxal and/or pyridoxine content of a 100 ml sulphuric acid sample test solution (7.2.1) should be between 2 and 25 µg.

7. PROCEDURE

7.1. *Remark*
Except where indicated all the analytical steps of the method must be performed without interruption. Duplicate determinations must be carried out on the same day.

7.2. *Disintegration*

7.2.1. *Sample test solution.* A total of 5–10 g of the untreated sample or 5–40 g of sample material which has been homogenised with sulphuric acid (6.2) is weighed to an accuracy of 0·01 g in a 100 ml volumetric flask, made up to 40 g with 0·1 M sulphuric acid (4.3), mixed with 10 drops of paraffin oil (4.9) in order to prevent foaming and heated in the autoclave

(5.4) for 30 min at 120 °C. After cooling to room temperature (20 °C), the solution is made up to 100 ml with water (paraffin oil is above the mark), thoroughly mixed and filtered through a dry pleated filter (5.11) (the first 10 ml of the filtrate being poured off). In order to remove ultrafine particles, an aliquot (approx. 3 ml) of the filtrate is passed through an ultrafilter (5.12) (= sample test solution).

> *Note 2:* Preserved, deep-frozen sample material (6.2) is brought carefully to room temperature before being weighed.

7.2.2. *External standard test solution.* For the external standard, 1–2 ml (= 10–20 µg) of the pyridoxamine, pyridoxal and pyridoxine standard solutions respectively (4.6.3; 4.7.3; 4.8.3) are pipetted into a 100 ml volumetric flask, made up to 40 ml with 0·1 M sulphuric acid (4.3) and autoclaved for 30 min at 120 °C. After cooling to room temperature (20 °C), they are made up to 100 ml with water (= standard test solution).

7.3. HPLC

A quantity of 50 µl of either the sample test solution (7.2.1) or the standard test solution (7.2.2) is added to the HPLC column, eluted with the mobile phase (4.4), and the fluorescent intensity of the separated pyridoxamine, pyridoxal and pyridoxine is measured. Standardisation and stability checks are performed using the standard test solution (7.2.2) at the beginning and during the course of a series of measurements (every sixth injection).

7.3.1. *HPLC conditions:*
Apparatus (5.13).
Stationary phase: Spherisorb RB ODS, 5 µm (5.20).
Mobile phase (eluant) (4.4).
Flow rate: 1·5 ml/min.
Volume to be applied: 50 µl of sample test solution (7.2.1) or standard test solution (7.2.2) (2–20 ng of pyridoxamine/pyridoxal/pyridoxine)
Detection: fluorimetry (5.15). Excitation wavelength, 290 nm; emission wavelength, 395 nm.
Retention time: pyridoxamine, 2·6 min; pyridoxal, 6·3 min; pyridoxine, 8·5 min.
Peak evaluation: over an integrated area using an integrator (5.14).

> *Note 3:* Since other peaks may occur in the natural matrix after pyridoxine,

elution must be continued until no further peaks appear. For most foodstuffs, elution lasts for approximately 30 min per injection.

7.3.2. *Chromatograms.* Figures 19, 20 and 21 show HPLC chromatograms relating to the determination of vitamin B_6 in standard and sample test solutions (baby food, beef, flour, lettuce).

8. EVALUATION

8.1. *Calculation*

The vitamin B_6 content of the analysis sample is calculated using the external standard. At both the start and end of a series of HPLC measurements, two or three 50 µl injections of the standard test solution (7.2.2) are performed and the average peak areas for pyridoxamine, pyridoxal and pyridoxine are calculated. The average peak areas of the sample test solutions (2–3 injections) are compared with those of the standard test solution; in this connection, account must be taken of the weighed portion, the removal of aliquots and dilutions.

The vitamin B_6 content of the sample (pyridoxamine, pyridoxal, pyridoxine) is calculated according to the following equations:

(a) *Sulphuric acid homogenate*

$$\frac{\text{PAM, PAL or POL}}{(\text{mg}/100\,\text{g sample})} = \frac{F_P C_S H \times 100}{F_S q E \times 1000}$$

(b) *Untreated sample material*

$$\frac{\text{PAM, PAL or POL}}{(\text{mg}/100\,\text{g sample})} = \frac{F_P C_S \times 100}{F_S E \times 1000}$$

Total vitamin B_6 (POL) content/100 g sample = (mg PAM × 1·006) + (mg PAL × 1·012) + (mg POL).

where: F_S = peak area for PAM, PAL or POL in the standard test solution (7.2.2); F_P = peak area for PAM, PAL or POL in the sample test solution (7.2.1); C_S = PAM, PAL or POL concentration in the standard test solution (µg/100 ml); H = total quantity of sulphuric acid homogenate (6.2) (g); q = weighed homogenate (7.2.1) (g); E = weighed portion for the homogenate (6.2.1) (g); 100 = conversion to 100 g of sample material; 1000 = conversion to mg of vitamin B_6.

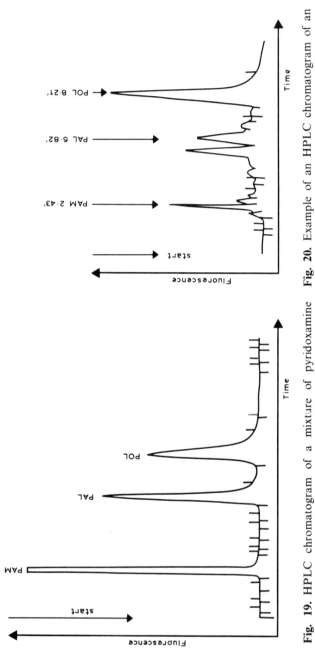

Fig. 19. HPLC chromatogram of a mixture of pyridoxamine (PAM), pyridoxal (PAL) and pyridoxine (POL). Standard test solution: 10 μg/100 ml.

Fig. 20. Example of an HPLC chromatogram of an extract from a vitamin B_6 enriched food item.

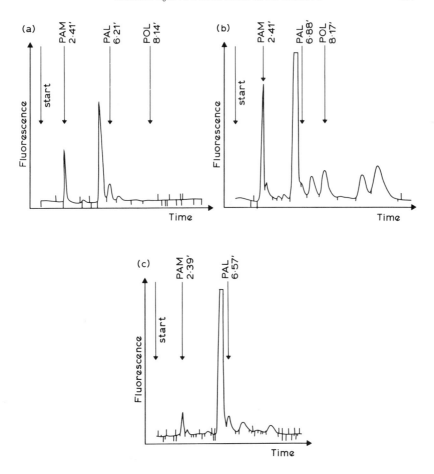

Fig. 21. Examples of HPLC chromatograms of extracts from unenriched foods: (a) beef; (b) rye flour; (c) lettuce.

8.1.1. *Range of linearity.* The linear proportionality of the peak area values of the sample test solutions compared with those of the standards are in the range 2–10 ng pyridoxamine, pyridoxal or pyridoxine/50 µl, corresponding to 4–20 µg/100 ml of the sample test solution.

8.1.2. *Detection limit.* Quantitative detection is possible down to 10 µg of pyridoxamine, 30 µg of pyridoxal and 30 µg of pyridoxine in a 100 g sample or 1–3 µg/100 ml of sample test solution.

8.2. Reliability of the Method

8.2.1. Recovery. The vitamin B_6 (pyridoxal, pyridoxamine and pyridoxine) recovery rates were between 94 and 102% depending on the foodstuff in question. Separate recovery measurements are unnecessary, since account is taken of all losses during the procedure by the use of the standard test solution.

8.2.2. Repeatability. The relative standard deviation in series of from two to five routine determinations in different foodstuffs (meat, milk, baby food, potatoes, vegetables) was $6\cdot5 \pm 3\cdot3\%$ for a vitamin B_6 content of 26–585 µg/100 g of sample.

8.2.3. Comparability. This was not determined for this method (no syndicate tests).

8.2.4. Comparison with other methods for the determination of vitamin B_6. Although comparative investigations were not conducted, a close correspondence was established between the analysis results obtained and the literature data.

9. ANALYSIS REPORT

The result of the determination is to be given with a reference to this method. Operations not mentioned in the described method must be indicated.

10. REFERENCES RELATED TO THE DESCRIBED PROCEDURE

10.1. Dr A. Bognár, Laboratory procedure, Institute for Nutritional Economics and Sociology in the Federal Research Institute for Nutrition, Garbenstr. 13, D-7000 Stuttgart 70, October 1983.
10.2. K. L. Lim, R. W. Young and J. A. Driskell; *J. Chromatog.* **188**, 285–8 (1980).
10.3. J. F. Gregory III, *J. Agric. Food Chem.,* **28**, 486–9 (1980).
10.4. J. F. Gregory III, *J. Food Sci.,* **45**, 84–6, 114 (1980).
10.5. J. T. Vanderslice, C. E. Maire, R. F. Doherty and G. R. Beecher, *J. Agric. Food Chem.,* **28**, 1145–9 (1980).

11. OTHER REFERENCES

High-pressure liquid chromatography is increasingly being used for the determination of vitamin B_6. It allows separate determination of the six vitamin B_6 vitamers. Microbiological methods however are still common. Microbiological methods are described in:

11.1. J.-P. Vuilleumier, H. P. Probst and G. Brubacher, Vitamine, Provitamine und Carotinoide. In: *Handbuch der Lebensmittelchemie II/2*, Ed. L. Acker et al., Springer Verlag, Berlin, 1967, pp. 833–6.
11.2. D. R. Osborne and P. Voogt, *The Analysis of Nutrients in Foods*, Academic Press, London, 1978, pp. 224–7.

The following HPLC methods described in the literature have already been put into practical use for food analysis and, where results from microbiological, GC or other methods are available for comparison, have proved to be both simpler and more precise. Rules for their use are generally specified in relation to particular substances.

11.3. J. T. Vanderslice, St. G. Brownlee, C. E. Maire, R. D. Reynolds and M. Polansky, *Am. J. Clin. Nutr.*, **37** (5) 867–71 (1983).
Forms of vitamin B_6 in human milk.
Method: HPLC.
Recovery: 90–106%.
Matrix: human milk.

11.4. J. T. Vanderslice, C. E. Maire and J. E. Yakupkovic, *J. Food Sci.*, **46**, 943–46 (1983).
Vitamin B_6 in RTE cereals: analysis by HPLC.
Method: HPLC.
Matrix: 24 different cereals, e.g. cornflakes, wheat flakes, rice crispies, grape nuts, cocoa puffs.

11.5. J. T. Vanderslice, C. E. Maire, R. F. Doherty and C. R. Beecher, *J. Agric. Food Chem.*, **28**, 1145–9 (1980).
Sulphosalicylic acid as an extraction agent for vitamin B_6 in food.
Method: SSA extraction; HPLC.
Recovery: 95–105% (pyridoxine phosphate 85%).
Matrix: pork, carp, hamburger, cereals.

11.6. K. L. Lim, *Diss. Abstr. Intern.*, **B 42** (10), 4021 (1982).
Comparison of HPLC, GLC and *Saccharomyces uvarum* methods for the determination of vitamin B_6 compounds.
Method: HPLC, GC–EC, *S. uvarum* assay; by comparison, HPLC is best, i.e. sensitivity, simplicity, precision.
Matrix: three foods.

11.7. K. L. Lim, R. W. Young and J. A. Driskell, *J. Chromatog.* **188**, 285–8 (1980).
Separation of vitamin B_6 components by HPLC.
Method: HPLC.
Matrix: homogenised and pasteurised skimmed milk.

11.8. J. F. Gregory III, D. B. Manley and J. R. Kirk, *J. Agric. Food Chem.*, **29**, 921–7 (1981).
Determination of vitamin B_6 in animal tissues by reverse-phase HPLC.
Method: $HClO_4$ extraction. HPLC with and without derivation of vitamin B_6 compounds.
Matrix: beef liver, beef, full milk, condensed milk, hamburger.

11.9. J. P. Gregory III, *J. Food Sci.*, **45**, 84–6, 114 (1980).
Bioavailability of vitamin B_6 in non-fat dry milk and a fortified rice breakfast cereal product.
Method: HPLC (individual vitamers); microbiological determination (total vitamin B_6); comparison.
Matrix: fat-free dried milk, breakfast cereals (rice).

11.10. J. P. Gregory III, *J. Agric. Food Chem.*, **28**, 486–9 (1980).
Comparison of HPLC and *S. uvarum* methods for the determination of vitamin B_6 in fortified breakfast cereals.
Method: HPLC (individual vitamers); microbiological determination (total vitamin B_6); comparison.
Matrix: torties, pebbles, crispies.

11.11. B. Hamaker, A. Kirksey and M. Borschel, *Fed. Proceedings*, **42** (5), Abstract No. 6077 (1983).
Distribution of B_6 vitamers in milk of pyridoxine supplemented mothers determined by HPLC.
Method: HPLC, five or six vitamin B_6 compounds found.
Matrix: human milk.

Finally, a reference which relates to a procedure using gas chromatography:

11.12. K. L. Lim, R. W. Young, J. K. Palmer and J. A. Driskell, *J. Chromatog.* **250**, 86–9 (1982).
Quantitative separation of B_6 vitamers in selected foods by a gas–liquid chromatographic system equipped with an electron-capture detector.
Method: GC–EC of naturally occurring derived vitamers.
Matrix: bread, milk, tinned peas.

12

Vitamin D in Margarine: HPLC Method

1. PURPOSE AND SCOPE

The method describes a procedure for the quantitative determination of vitamin D in margarine. It determines added vitamin D_2 (ergocalciferol) and D_3 (cholecalciferol) in products naturally free from vitamin D. The relevant previtamins D_2 and D_3 and, if present, other isomers and transformation products are not detected with this method. Vitamins D_2 and D_3 are not separated by this procedure. The method is only applicable to margarine.

> *Note 1:* In initial tests the authors (Ref. 10.2) succeeded with their laboratory method in also determining vitamin D in milk (300 IUD/litre) and cooking fat (9000 IUD/kg). The method described has only been used for margarine with at least 750 IUD/100 g (Swiss margarine).

Vitamin D contents from 17·5 µg (= 700 IU)/100 g of sample can be determined quantitatively; contents above 1·5 µg (= 60 IU)/100 g can be estimated qualitatively.

2. DEFINITION

The content of vitamin D is understood to be the ergocalciferol and/or cholecalciferol content determined by the procedure described here: it is given in IUD/100 g.

1 IUD corresponds to 0·025 µg (= 25 ng) vitamin D_2 or vitamin D_3.
1 µg vitamin D corresponds to 40 IUD.
1 ng vitamin D corresponds to 0·04 IUD.

3. BRIEF DESCRIPTION (PRINCIPLE OF THE METHOD)

After alkaline saponification of the sample, and exhaustive extraction of the non-saponifiables with diethyl ether, the extract residue is dissolved in methanol, and an aliquot of the extract is precleaned in a reverse-phase semi-preparative HPLC column. After collection of the vitamin D–containing eluate fraction and its transfer into n-hexane or isooctane, vitamin D is separated on a straight-phase analytical HPLC column. The measurement is carried out at 264 nm and the peak areas or heights are evaluated against an external vitamin D standard.

4. CHEMICALS

Remark: Unless otherwise specified, AR grade chemicals are to be used. Distilled water from glass vessels or water of corresponding purity must be used.

4.1. Potassium hydroxide, pellets, extremely pure (content 85%), e.g. Merck.
4.2. Vitamin D_3 (cholecalciferol), crystalline for biochemical purposes; content 40 million IU/g, e.g. Merck or vitamin D_3, Fluka, very pure, crystalline, over 98%.
4.3. Vitamin D_2 (ergocalciferol), crystalline, for biochemical purposes; content 40 million IU/g, e.g. Merck or vitamin D_2, Fluka, extremely pure, crystalline, over 99%.
4.4. Sodium chloride (common salt), crystalline, pure, Ph. Eur., e.g. Fluka, or sodium chloride, crystalline, AR, e.g. Merck.
4.5. BHT (2,6-di-*tert*-butyl-4-methylphenol) for synthesis, e.g. Merck.
4.6. n-Hexane, e.g. Merck.
4.7. Isooctane, e.g. Merck.
4.8. 1,4-Dioxane, AR, stabilised with approximately 25 ppm BHT, e.g. Merck.
4.9. Methanol, e.g. Merck.
4.10. Ethanol, absolute, e.g. Merck.
4.11. Diethyl ether, AR, stablished with BHT, e.g. Merck.
4.12. Nitrogen, oxygen-free, 99·9% (v/v).
4.13. Solutions.
    4.13.1. Potassium hydroxide solution, 50% (w/v) aqueous; dissolve 50 g KOH (4.1) in water, with cooling, to make 100 ml.

4.13.2. Sodium chloride solution, 10% (w/v) aqueous; dissolve 10 g NaCl (4.4) in water, to make 100 ml.

4.13.3. *Vitamin D standard solution.* Approximately 25 mg vitamin D_3 (4.2) or vitamin D_2 (4.3), weighed to an accuracy of 0·01 mg, are dissolved in isooctane (4.5) to make 100 ml with the addition of a few crystals of BHT (4.7) (= stock solution with approx. 250 μg/ml). The solution can be kept for up to six months at +4 °C. A volume of 1·0 ml stock solution is diluted with isooctane (4.7) to make 100 ml (vitamin D standard solution with approx. 2·5 μg = 100 IU/ml). The solution can be kept for 14 days at +4 °C.

5. APPARATUS AND AUXILIARY EQUIPMENT

5.1. Standard basic laboratory equipment including refrigerator and deep freezer.

5.2. 250 ml ground-glass round flask with reflux condenser and N_2 supply to the condenser.

5.3. Water or steam bath; electrical heating bath.

5.4. Two separating funnels, Squibb, 500 ml.

5.5. HPLC column (semi-preparative): LiChrosorb RP–8, 7 μm, 250 × 10 mm (Hibar; Merck) for reverse-phase chromatography.

5.6. HPLC column (analytical): LiChrosorb Si 60, 5 μm, 250 × 4 mm (Hibar; Merck) for straight-phase chromatography.

5.7. HPLC apparatus: commercially available pump (e.g. Altex).
   Injection valve, e.g. Rheodyne with injection volumes of up to 500 μl (manual).
   Detector: commercially available, 264 nm.
   Recorder: commercially available.
   } for HPLC precleaning (7.4)

   Pump: as above.
   Injection valve, e.g. Rheodyne with injection volumes up to 100 μl.
   Detector: as above.
   Recorder: commercially available.
   } for HPLC analysis (7.7)

Work is to be carried out on two separate HPLC lines.

6. SAMPLE

6.1. Sampling

The sample to be taken must be representative of the margarine to be analysed with regard to proportions and composition. About 100 g of sample should be available.

6.2. Sampling Method and Preparation of the Sample

When margarine is sampled, care should be taken to see that the date of production or sell-by date, production number, type of packaging and type of margarine are recorded. Should the margarine in the pack not have a standard appearance (partly melted, discoloration), it should be melted carefully at 40 °C and mixed before taking a weighed portion in the hot state. In other cases it is to be assumed that the distribution of vitamin D in the (solid) margarine is uniform.

6.2.1. *Weighed portion.* Approximately 20 g of margarine is to be weighed out for the range 100 to 750 IUD/100 g.

7. PROCEDURE

7.1. Remark

The procedure is to be carried out without major interruption as a true double determination (two weighed samples). Possible interruption points are indicated. Direct sunlight should be avoided. All operations should be carried out under nitrogen.

7.2. Saponification

Approximately 20 g of the margarine sample, prepared in accordance with method 6.2 and weighed out to an accuracy of 0·1 g, is mixed in a 250 ml round flask with reflux condenser (5.2) with 70 ml ethanol (4.10), 50 mg BHT (4.5) and 20 ml of 50% potassium hydroxide solution (4.13.1). Any air present in the system is removed with nitrogen (4.12) and the mixture is heated to boiling for 20 min with suitable agitation in a water bath or heating bath (5.3). The flask is cooled to approx. 40 °C and the contents transferred to a 500 ml Squibb separating funnel (5.4) with subsequent rinsing of the flask with 100 ml water (=saponification solution).

7.3. Extraction

The saponification solution (7.2) is immediately diluted with 120 ml diethyl ether (4.11); there follows 30 s intensive shaking and then the complete separation of the two phases. The lower phase (saponification solution) is led off into a second separating funnel and extracted in similar fashion with 120 ml diethyl ether (4.11). The combined diethyl ether extracts are then washed, until neutral (against phenolphthalein; pH paper), with the following:

(1) 100 ml of 10% sodium chloride solution (4.13.2).
(2) 100 ml water.
(3) 100 ml water containing 10% ethanol (4.10).
(4) 100 ml water.

After the addition of approx. 20 mg BHT (4.5), the diethyl ether solution is evaporated in a rotary evaporator under a partial vacuum on the water bath, at a maximum of 40 °C bath temperature, until almost dry; after the addition of 10 ml hexane (4.6)–ethanol (4.10) mixture (1:1), the residue is made free from water by means of azeotropic evaporation, transferred with methanol (4.9) into a 2·0 ml graduated flask and made up to the mark (=raw extract).

Note 2: The raw extract can be kept overnight at +4 °C. Turbid extracts may be cleaned by centrifugation.

7.4. Purification of Extract by Means of Semi-preparative HPLC (Reverse Phase)

A 0·5 ml aliquot of the raw extract (7.3) is precleaned by reverse-phase HPLC under the following conditions:

Apparatus (5.7).
Stationary phase: column (5.5).
Mobile phase: methanol (4.9)—water (90:10).
Quantity: 500 µl (by loop).
Flow rate: 2·0 ml/min.
Temperature: room temperature.
Detection wavelength: 264 nm.
Detector sensitivity: 0·1 AUFS.
Duration of analysis: at least 60 min up to a second injection.
Recorder speed: 5 mm/min.

The fraction (approx. 16 ml) is collected between the 24th and the 32nd minute, evaporated to the dry state in a 50 ml flask and dried azeotropically as described in procedure 7.3. The residue is dissolved in exactly 0·5 ml isooctane (4.7) (= pure extract).

Note 3: The pure extract can, if necessary, be kept overnight at +4 °C.

7.5. Sample Test Solution

The pure extract (7.4) also serves as the sample test solution.

7.6. Standard Test Solution

The vitamin D standard solution (4.13.3) also serves as the standard test solution. It is used for the external calibration of the HPLC line and for checking the stability of the UV measurements; it is not subjected to full analysis.

7.7. HPLC

7.7.1. Calibration of the analytical HPLC line (straight phase) and plotting of the standard test curve. Before a series of measurements is started the HPLC apparatus is calibrated externally with a double measurement using 100 µl of the vitamin D standard test solution (4.13.3), corresponding to 10 IUD, and with further dilutions of this solution with isooctane (4.7), which in 100 µl contains, respectively, 7·5, 5·0, 2·5, 1·0 and 0·5 IUD. The signals are evaluated by means of peak areas (integrator) or peak heights (chromatogram) and the mean values are used for plotting the standard test curve. Chromatography is carried out isocratically at room temperature under the following conditions:

Apparatus: (5.7).
Stationary phase: column (5.6).
Mobile phase: dioxane (4.8)–*n*-hexane (4.6) (7:93).
Quantity: 100 µl (by loop).
Flow rate: 1·2 ml/min.
Temperature: room temperature.
Detection wavelength: 264 nm.
Detector sensitivity: 0·02 to 0·04 AUFS.
Retention time for vitamin D: approx. 16·5 min.
Recorder speed: 5 mm/min.
Duration of analysis: 30 min.

During the series of measurements, the calibration should be checked repeatedly (e.g. every fifth injection) with 2·5 IUD/100 µl from the above series of dilutions.

7.7.2. *Measurement of the sample.* On two occasions 100 µl of the sample test solution (7.5) are injected as described in section 7.7.1 and the peak signal (mean value) evaluated from the standard test curve (7.7.1)

7.7.3. *Chromatograms.* Figures 22 and 23 show HPLC chromatograms for margarine with respect to extract purification (7.4), measurements of samples (7.7.2) and single standards (7.7.1).

8. EVALUATION

8.1. *Calculation*

The peak heights or areas of the samples measured (7.7.2) are compared with those of the standard test curve (7.7.1); weighing, size of aliquots and any dilutions should be taken into account. The following equation should be used:

$$\text{Vitamin D content (IU/100 g)} = \frac{H_p \times S \times 100 \times F}{H_s \times E}$$

(To give the value in µg vitamin D/100 g, divide by 40.) In the equation: H_p = peak height in HPLC chromatogram for 100 µl of sample test solution (7.5) (mm); H_S = peak height in HPLC chromatogram for 100 µl of standard test solution (7.6) (mm); S = quantity of vitamin D in 100 µl injected standard test solution (7.6) (IUD); E = weighed portion of margarine (g); F = factor—relates to dilution and aliquots. ($F = 40$— relates to the quantities given in sections 7.2–7.7.2).

8.1.1. *Range of linearity.* The standard test curve (7.7.1) is linear in the range 0·5 to 10 IUD.

8.1.2. *Detection Limit.* A quantity of 17 µg vitamin D_2 (vitamin D_3) (= 700 IUD in 100 g of sample or 0·125 µg vitamin D (= 5 IUD) in 1 ml sample test solution) can be quantitatively determined. An amount of 1·5 µg vitamin D (= 60 IUD) is still qualitatively detectable in 100 g of sample.

8.2. *Reliability of the Method*

8.2.1. *Recovery.* Recovery tests were carried out as described with an addition of 20 000 IU of vitamin D_3/kg in vitamin D_3-free margarine

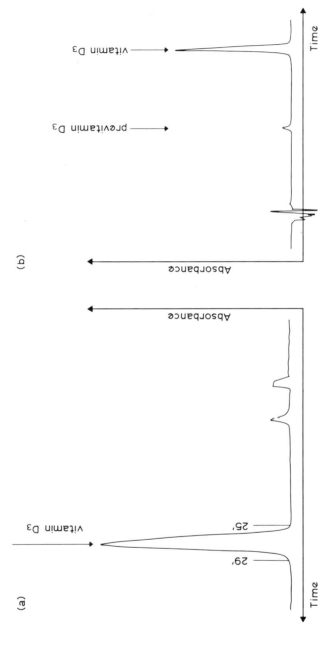

Fig. 22. HPLC chromatograms of vitamin D_3 standard solution. (a) Precleaning conditions: LiChrosorb RP-8, 7 µm; methanol: water (90:10); 264 nm; 0·1 AUFS; 2·0 ml/min. (b) Determination conditions: LiChrosorb Si 60, 5 µm; hexane: dioxane (93:7); 0·04 AUFS; 264 nm; 1·2 ml/min.

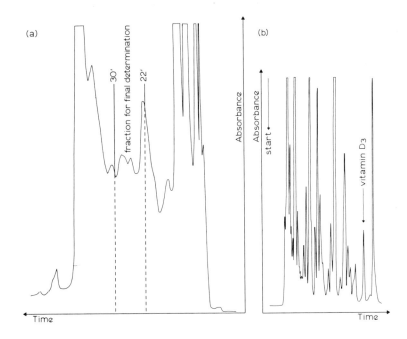

Fig. 23. HPLC chromatogram of an extract from margarine containing 750 IU vitamin D_3/100 g. (a) Precleaning of an extract from 5 g of margarine—for conditions, see Fig. 22(a). (b) Determination of vitamin D_3, 100 μl of fraction 22′–30′—for conditions, see Fig. 22(b).

using full analysis, i.e. 20 g margarine was enriched with 400 IU of vitamin D_3 (in hexane solution). The recovery rate with five determinations was 93% with a coefficient of variation of 4·5%. The recovery value for the two HPLC stages (7.4 and 7.7) was 97% with a coefficient of variation of 2% (five determinations).

With regard to the recovery values it must be taken into account that any previtamin D_3 which may form during saponification is separated during the second HPLC stage (7.7) and will not be detected. Process losses and losses which are difficult to calculate because of previtamin D_3 formation during the course of the analysis require a correction estimated at 10% for the measured result.

8.2.2. *Repeatability.* This was not determined.

8.2.3. *Comparability.* This was not determined since there are no syndicate tests.

8.2.4. *Comparison with other methods for determining vitamin D.* No comparisons for margarine are available.

9. ANALYSIS REPORT

The result of the determination is to be given with a reference to this method. Any operations which are not mentioned in the method should be listed.

10. REFERENCES

> *Remark:* The method described is based on details of the Laboratory methods for the determination of vitamin D in margarine used in the Department for Vitamin and Nutrition Research, Hoffmann-La Roche, Basel (10.1) and the Swiss Vitamin Institute, Basel (10.2), as at October 1983.

The other literature on vitamin D determination in margarine covers only those methods which (a) were used on foodstuffs, (b) only or also apply to margarine and (c) use the HPLC method. The method described contains the essential elements from publications from the period 1980–82 (Refs. 10.3–10.6).

There are as yet no COST 91 methods for vitamin D determination in milk (and milk products) and eggs; the relevant literature from the period 1977–82—milk (Refs. 10.7–10.8), eggs (Refs 10.19–10.21)—are also given and relate exclusively to HPLC methods. The methods marked with an asterisk are those most worth recommending.

The literature on vitamin D determination in dietetic foods, children's foods, etc., is not included in the references.

10.1. U. Manz and K. Philipp, Abteilung VF der F. Hoffmann-La Roche & Co. AG; Basel, October 1983.
10.2. P. Walter and F. Brawand, Schweizerisches Vitamininstitut, Basel, October 1983.

Vitamin D in Margarine
10.3. P. J. van Niekerk and S.C.C. Smit, *J. Am. Oil Chem. Soc.*, **57** (12), 417–21 (1980).

10.4. W. Müller-Mulot and G. Rohrer, *Fette-Seifen-Anstrichmittel*, **84**(9), 354–8 (1982).
10.5. J. N. Thompson, G. Hatina and W. B. Maxwell, *J. Assoc. Off. Anal. Chem.*, **65**(3), 624–31 (1982).
10.6. F. Zonta, B. Stancher and J. Bielawny, *J. Chromatog.* **246**, 105–112 (1982).

Vitamin D in Milk

10.7. J. N. Thompson, W. B. Maxwell and M. L'Abbé, *J. Assoc. Off. Anal. Chem.*, **60**(5), 998–1002 (1977).
10.8. S. K. Henderson and A. F. Wickroski, *J. Assoc. Off. Anal. Chem.*, **61**(5), 1130–4 (1978).
*10.9. S. K. Henderson and L. A. McLean, *J. Assoc. Off. Anal. Chem.*, **62**(6), 1358–60 (1979).
10.10. A. Adachi and T. Kobayashi, *J. Nutr. Sci. Vitaminol.*, **25**, 67–78 (1979).
*10.11. T. Okano, A. Takeuchi and T. Kobayashi, *J. Nutr. Sci. Vitaminol.*, **27**(6) 539–50 (1981).
10.12. B. W. Hollis, B. A. Roos, H. H. Draper and P. W. Lambert, *J. Nutr.*, **111**(7), 1240–8 (1981).
10.13. B. Borsje, E. de Vries, J. Zeeman and F. J. Mulder, *J. Assoc. Off. Anal. Chem.*, **65**(5), 1225–7 (1982).
10.14. E. de Vries and B. Borsje, *J. Assoc. Off. Anal. Chem.*, **65**(5), 1228–34 (1982).
10.15. W. Müller–Mulot, B. Hippin and G. Rohrer, Hoffmann-La Roche AG, Grenzach, Rapport v. 18.10.1982.
10.16. R. A. Wiggins, E. S. Zai and I. Lumley, *Chromatog. Sci.*, **20**, 327–41 (1982).
10.17. J. F. Muniz, C. T. Wehr and H. M. Wehr, *J. Assoc. Off. Anal. Chem.*, **65**(4), 791–7 (1982).
*10.18. L. Mouillet, F. M. Luquet, M. F. Gagnepain and Y. Sorgue, *Le Lait*, **62**, 44–54 (1982).

Vitamin D in Eggs

*10.19. P. A. Jackson, C. J. Shelton and P. J. Frier, *Analyst*, **107**, 1363–9 (1982).
10.20. P. A. Jackson, *Food*, 12–18 (December 1982).
*10.21. L. M. Sivell, R. W. Wenlock and P. A. Jackson, *Human Nutrition: Applied Nutrition*, **36A**, 430–7 (1982).

Vitamin D in Fat, Oil and Margarine

10.22. M. Rychener and P. Walter, *Mitt. Gebiete Lebensm. Hyg.*, **76**, in press (1985).

PART IV

ANNEX

13
Niacin

The main problem in niacin determination is that niacin occurs in natural products in the form of free nicotinic acid and nicotinic acid amide (nicotinamide) on the one hand and in bound forms on the other. It is said that bound forms, which can be transformed into free niacin by acid hydrolysis, are fully bioavailable for man, whereas bound forms, which can only be transformed by alkaline hydrolysis, have no nutritive value. Acid-hydrolysable forms occur in plant and animals, e.g. nicotinamide adenine dinucleotide (NAD) and its phosphate (NADP), whereas forms which can only be hydrolysed by alkali occur mainly in plants. Thus, if the nutritive value is to be determined, an acid hydrolysis is preferable, whereas if the total amount of niacin is to be determined, an alkaline procedure has to be chosen.

Niacin poses few technological problems, because the free forms are very resistant to heat, light, oxidation, acids and alkalis. However, even if only the acid-hydrolysable part is determined it is difficult to estimate the nutritive value of a food item, because for this purpose it is necessary to know also its tryptophan content—tryptophan acts as a niacin precursor in the human body.

Microbiological methods for determining niacin are still common. In general, microorganisms which behave similarly to man, with regard to bioavailability of the various niacin forms, are used. For natural products, depending on the conditions of hydrolysis, the results of a microbiological assay may differ appreciably from the results of a chemical method.

REFERENCES FOR MICROBIOLOGICAL DETERMINATION OF NIACIN

1. D. R. Osborne and P. Voogt, *The Analysis of Nutrients in Foods*, Academic Press, London, 1978, pp. 221–3.

See also:
2. J.-P. Vuilleumier, H. P. Probst and G. Brubacher, Vitamine, Provitamine und Carotinoide. In: *Handbuch der Lebensmittelchemie II/2*, Ed. L. Acker *et al.*, Springer Verlag, Berlin, 1967, p. 832.

RECENT REFERENCES FOR CHEMICAL DETERMINATION OF NIACIN

3. *Niacin in Meat and Meat Products*
3.1. K. Yoshida, Y. Yamamoto and M. Fujiwara, *Shokuhin Eiseigaku Zasshi*, **23**(6), 428–33 (1982).
A simple analytical method for niacin and nicotinamide in foods by HPLC.
Method: HPLC reverse-phase; ODS column, ion-pair.
Mobile phase: methanol–sodium acetate solution with tetrabutylammonium hydroxide.
Detection Limit: 1 mg niacin/100 g.
Recovery: >90%.
Matrix: beef, fish.
3.2. Y. Kitayama, M. Inoúe, K. Tamase, M. Imou, A. Hasuike, M. Sasaki and K. Tanigawa, *Eiyo to Shokuryo*, **35**(2), 121–4 (1982).
Determination of free niacin in foods by ion-pair HPLC.
Method: HPLC, reverse-phase; Lichrosorb RP8 column, 25 cm.
Mobile phase: solution with tetrabutylammonium bromide.
Detection limit: 0·5 mg/100 g.
Recovery: 96–102% (added).
UV: 260 µm.
Matrix: pork, beef, chicken, fish.

4. *Niacin in Cereals*
4.1. R. B. Roy and J. J. Merten, *J. Assoc. Off. Anal. Chem.*, **66**(2), 291–6 (1983).
Evaluation of urea–acid system as medium of extraction for the B-group vitamins. Part II. Simplified semi-automated chemical analysis for niacin and niacinamide in cereal products.
Method: disintegration in HCl–urea solution, 30 min, 15 psi (autoclave); colorimetric determination by König–Zincke colour reaction (cyanogen bromide, sulphanilic acid), 480 nm.
Recovery: 99·4–99·8% (bread and flour enriched).
Matrix: wheatgerm, rolled oats, flour, rice, bran, bread, also hot dog rolls and hamburger rolls.
4.2. AOAC, *Official Methods of Analysis*, 13th edn., AOAC, Washington, DC, 1980, pp. 743–6.
Method: disintegration in calcium hydroxide; colorimetric determination by König–Zincke colour reaction (cyanogen bromide, sulphanilic acid).

5. Niacin in Fruit
5.1. K. Kral, *Fresenius' Z. Anal. Chem.*, **314**(5), 479–82 (1983). Determination of nicotinic acid in fruit juices by HPLC with electrochemical detection on a stationary dropping-mercury electrode.
Method: Disintegration in H_2SO_4, 30 min, 220 °C (autoclave);
HPLC, reverse-phase, Nucleosil 5 NH_2 column, 30 cm.
Mobile phase: methanol–sodium acetate solution (pH 4·6).
Detection: (a) UV, 260 nm, with detection limit 9 ng (for standard solutions);
(b) electrochemical detection with detection limit 40 ng (for complex matrix).
Determined: nicotinic acid + nicotinamide + derivatives.
Coefficient of variation: 2·3–3·7%.
Matrix: pear, orange and apple juice.

6. Niacin in Various Foodstuffs
6.1. D.C. Egberg, *J. Assoc. Off. Anal. Chem.*, **62**(5), 1027–30 (1979).
Semiautomated method for niacin and niacinamide in food products, collaborative study.
Method: Disintegration in calcium hydroxide, 2 h, 120 °C (autoclave). Colorimetric determination by König–Zincke colour reaction (cyanogen bromide, sulphanilic acid) 470 nm. Based on the AACC methods (American Association of Cereal Chemists, Approved Methods), 1976, St Paul, MN, Method 86–51 and D. C. Egberg, R. H. Potter and G. R. Honold, *J. Agr. Food Chem.*, **22**(2), 323–6 (1974).
Recovery: 94–100%
Coefficient of variation: 1·5%   } Egberg et al., 1974.
Detection limit: 0·8 mg/100 g
Matrix: soya flour, yeast, dried potatoes, RTE cereals, noodles, rice, wheat flour, breadcrumbs, pudding, cocoa, etc.
6.2. AOAC, *Official Methods of Analysis*, 13th edn, AOAC, Washington, DC, 1980, pp. 743–6.
Method: disintegration in H_2SO_4 (with exception of cereals, see Ref. 4.2).

14

Folacin in Foodstuffs

The term folacin or folates comprises a group of compounds having nutritional properties and chemical structures similar to those of folic acid (pteroylglutamic acid). Folic acid itself does not occur naturally in food, but is used for enrichment. Most of the naturally occurring folates are derivatives of 5,6,7,8-tetrahydrofolic acid and exist in monoglutamate and polyglutamate forms (see Fig. 9).

The various folacin-active compounds vary widely in bioavailability. Free folic acid seems to be the most active compound, whereas polyglutamates are several times less active. It is therefore essential to determine each compound separately if the nutritive value of a food item is to be evaluated; this is not possible at the moment with commonly used methods. Also, the determination of total folacin by converting all folacin active compounds into free folic acid is still an unsolved problem, since some compounds are heat stable whereas others are rapidly destroyed by heat, and some compounds are destroyed by acid whereas others are quite stable at low pH. Free folic acid itself is destroyed in neutral or alkaline solution by heat and may be oxidised by oxygen at higher pH.

At the moment the only practical way of assaying folic acid active compounds in foods consists of microbiological methods, the most recommendable of which was described by A. E. Bender and N. I. Nik-Daud at the final COST 91 seminar at Athens on 14–18 November 1983 (Ref. 1.1).

There are many methods for separating the folacin-active compounds by HPLC in specified substrates and it is hoped that in a few years time a generally applicable method will be elaborated.

REFERENCES

1. Microbiological Methods

1.1. A. E. Bender and N. I. Nik-Daud, *Folic acid: Assay and Stability*, Proceedings of the final COST 91 seminar in Athens, 14–18 November, 1983. In: *Thermal Processing and Quality of Foods*, Ed. P. Zeuthen *et al.*, Elsevier Applied Science Publishers, London pp. 880–4.

2. HPLC Methods

2.1. C. K. Clifford and A. J. Clifford, *J. Assoc. Off. Anal. Chem.*, **60**, 1248–51 (1977).
HPLC analysis of food for folates.
Determination method:HPLC.

2.2. D. R. Briggs, G. P. Jones and P. Sae-Eüng, *Proc. Nutr. Soc. Australia*, **6**, 149 (1981).
The determination of folacin in food.
Determination method:HPLC.

2.3. D. R. Briggs, G. P. Jones and P. Sae-Eüng, *J. Chromatog.* **246**(1), 165–8 (1982).
Isocratic separation of food folacin by HPLC.
Determination method: HPLC.
Recommended for folacin determination in foodstuffs.

2.4. B. P. F. Day and J. F. Gregory III, *J. Agric. Food Chem.*, **29**(2), 374–7 (1981).
Determination of folacin derivatives in selected foods by HPLC.
Determination method: HPLC.
The folacin content of cow's liver, enriched breakfast cereals and baby food was determined.

2.5. J. P. Gregory III, B. P. F. Day and K. A. Ristow, *J. Food. Sci.*, **47**(5), 1568–71 (1982).
Comparison of HPLC, radiometric and *Lactobacillus casei* methods for the determination of folacin in selected food.
Determination method: HPLC.
The folacin concentration in raw cabbage, an enriched cereal and baby food was determined.
The HPLC method was clearly superior.

2.6. K. Hoppner and B. Lampi, *J. Liq. Chromatog.*, **5**(5), 953–66 (1982).
The determination of folic acid (pteroylmonoglutamic acid) in fortified products by HPLC.
Determination method: HPLC.
Fortified baby foods based on milk and soya protein were analysed.

2.7. M. C. Thomas, *Diss. Abstr. Intern.*, **41**(7), 2545–6-B (Jan. 1981).
An approach to the chemical assay of folates.
Determination method: HPLC.
Model experiments.

Index

Accuracy, definition of, 17
Allyl thiamine disulphide, 7
Aluminium oxide, deactivated,
　preparation of, 34–5, 40, 46
Amberlite CG 50 I cation-exchanger,
　51, 52, 59
Aminex-A 14 anion-exchanger, 68–9,
　71
Apparatus listed, 3–4
　carotene assays, 35, 41, 46
　vitamin A assay, 25
　vitamin B_1 thiochrome method,
　　54–6
　vitamin B_2 HPLC method, 121
　vitamin B_6 HPLC method, 131–2
　vitamin C
　　HPLC method, 68–9
　　modified Deutsch and Weeks
　　　fluorimetric method, 78–9
　　Sephadex method, 88–9
　vitamin D HPLC method, 143
　vitamin E HPLC methods, 99,
　　109
Ascorbic acid (ASC), 10
　oxidation of, 76, 80–1, 85–6, 91
　see also Vitamin C
Ascorbyl palmitate, 10

Barley, tocopherols in, 104
Beef, vitamin B_6 compounds in, 137
Bender–Nik-Daud microbiological
　method, folacin compounds
　assay, 158, 159
p-Benzoquinone, ascorbic acid
　oxidised by, 85–6, 91
Bias, analytical error, 16
Biological activity effects, 13

Biotin, methods/references omitted
　for, 5
Bondapak C_{18} columns, 25, 29, 121,
　124
Bound niacin, 13, 155
Bound riboflavin, 123
British Anti-Lewisite (BAL), 66, 67,
　70, 85, 86, 90–1
Bundesanstalt method, vitamin B_1
　thiochrome method compared
　with, 64

Carcinogens, 1,4-dioxane, 108
Carotene, recommended methods,
　33–49
　beverages, 45–9
　complex foodstuffs, 34–9
　natural fruits and vegetables, 39–45
Carr–Price method, vitamin A HPLC
　method compared with, 31
Cereals, niacin in, 13, 156, 157
Chemicals listed, 3
　carotene assays, 34–5, 40, 45–6
　vitamin A assay, 24–5
　vitamin B_1 thiochrome method,
　　52–4
　vitamin B_2 HPLC method, 120
　vitamin B_6 HPLC method, 130–1
　vitamin C
　　HPLC method, 67–8
　　modified Deutsch and Weeks
　　　fluorimetric method, 77–8
　　Sephadex method, 86–8
　vitamin D HPLC method, 142–3
　vitamin E HPLC methods, 98, 108
Cholecalciferol, 11
　determination of, 141–50

Chromatographic techniques
  carotene assay, 37–8, 43, 47
  see also High-pressure liquid chromatography
Clara-Diastase suspension, 53, 58, 120, 122
Collaboration, analytical trials, 19
Colorimetry
  vitamin C assay, 93–4
Comparability, definition of, 17
Complex foodstuffs
  carotene in, 34–9
  tocopherols in, 104
Comprehensive methods, impracticability of, 5–6
COST 91, v
  project leader, v

Dehydroascorbic acid (DASC), 10
  ascorbic acid oxidised to, 76, 80–1
  reduction of, 66, 70, 85, 90–1
  vitamin C measured as, 76, 80–1
  see also Vitamin C
Dehydroretinaldehyde, 6
Dehydroretinol, 6
Detection limits
  carotene, 33, 38, 44, 48
  general discussion of, 18
  α-tocopherol, 97, 105
  vitamin A, 23, 30
  vitamin B_1, 51, 63
  vitamin B_2, 119, 125
  vitamin B_6, 129, 137
  vitamin C, 66, 73, 76, 83, 85, 95
  vitamin D, 141, 147
Deutsch and Weeks extraction medium, vitamin C assay, 78
Deutsch and Weeks fluorimetric method, vitamin C assay, modified method, 76–84
Dextrin-containing foods, vitamin C assay for, 93
DGF method, vitamin E assay, 107–15
2,6-Di-*tert*-butyl-4-methylphenol (BHT), 98, 99, 142, 144

2,6-Di-*tert*-butyl-4-methylphenol—*contd.*
  1,4-dioxane stabilised with, 108, 142
  HPLC peaks, 103–4
2,3-Dimercapto-1-propanol (BAL), 66, 67, 70, 85, 86, 90–1
1,4-Dioxane, 108, 113, 142
Disintegration techniques
  vitamin B_1 assay, 57
  vitamin B_2 assay, 122
  vitamin B_6 HPLC method, 133–4
  vitamin C methods, 70, 80, 90

Eggs, vitamin D in, 151
Enzymatic digestion, 58, 122–3
Ergocalciferol, 11
  determination of, 141–50
Errors
  statistical, 16, 17
  systematic, 16, 17
  types of, 16
Erythorbic acid (isoascorbic acid), ascorbic acid not distinguished from, 66
N-Ethylmaleimide (NEM), 85, 90
Evaluation
  carotene methods, 38, 44, 48
  general discussion of, 4, 16–19
  selection criteria stated, 4
  tocopherols determination, 102, 114–15
  vitamin A HPLC method, 30–1
  vitamin B_1 thiochrome method, 62–3
  vitamin B_2 HPLC method, 125–7
  vitamin B_6 HPLC method, 135
  vitamin C
    HPLC method, 73
    modified Deutsch and Weeks fluorimetric method, 83
    Sephadex method, 94
  vitamin D compounds, 147
  vitamin E HPLC methods, 102, 114–15
Expert group, listed, ix–x

Extraction techniques
  carotene assay, 36–7, 42–3, 47
  vitamin A method, 27–8
  vitamin B_1 method, 57–8
  vitamin B_2 method, 122
  vitamin B_6 method, 133–4
  vitamin C methods, 70, 80, 90
  vitamin D assay, 145
  vitamin E assay, 100

Flavin adenine dinucleotide (FAD), 8
Flavin mononucleotide (FMN), 8
Fluorimetric detection
  vitamin B_2 HPLC method, 124
  vitamin B_6 assay, 134
  vitamin E group compounds, 97, 102, 107, 113
Fluorimetric measurements, thiochrome, 62
Fluorimetry, ascorbic acid quinoxaline derivative, 82
Folacin group
  HPLC methods for, 159
  microbiological methods for, 158–9
  references only given for, 4, 158–9
  vitamers listed for, 12
Folic acid, 12, 158
Food composition tables
  accuracy in, 18
  vitamin content for, 5, 12–13
Food processing, vitamins affected during, 5, 18
Fruit, niacin in, 157

High-pressure liquid chromatography (HPLC)
  folacin-active compounds (references only), 159
  reverse-phase, 25, 29–30, 119, 124, 142, 143, 145–6
  stationary phases
    Aminex-A 14, 68–9, 71
    Bondapak C_{18}, 25, 29, 121, 124
    LiChrosorb RP-8, 143, 148
    LiChrosorb RP-18, 121, 124

High-pressure liquid chromatography (HPLC)—contd.
  stationary phases—contd.
    LiChrosorb Si 60, 99, 101, 109, 113, 143, 148
    Spherisorb RB ODS, 130, 132, 134
    Supercosil C_8, 25, 29
  straight phase, 98, 101–2, 109, 113, 142, 143, 146–7
  vitamin A compounds, 25, 29–30
  vitamin B_2 assay, 121, 124
  vitamin B_6, 130, 132, 134–5, 136–7
    comparison studies for, 139–40
  vitamin C, 71–2
  vitamin D compounds, 142, 143, 146–7, 148–9
  vitamin E compounds, 98, 101–2, 103–4, 107, 112–14

Ion-exchangers
  Amberlite CG 50 I, 51, 52, 59
  Aminex-A 14, 68–9, 71
  Sephadex DEAE A-25, 86, 88, 91
IU equivalents
  vitamin A, 23, 24–25
  vitamin D, 141

Labelling legislation, vitamin content for, 5, 13
Legal requirements, vitamin determination for, 5, 13
Lettuce, vitamin B_6 compounds in, 137
LiChrosorb RP-8 columns, 143, 148
LiChrosorb RP-18 columns, 121, 124
LiChrosorb Si 60 columns, 99, 101, 143, 148
Lycopin, carotenes separated from, 38, 47

Manz–Philipp HPLC method, α-tocopherol assay, 105
Margarine
  HPLC chromatogram of, 149
  vitamin D compounds in, 141–50

INDEX

Meat products
  niacin in, 156
  vitamin B_6 compounds in, 137
Metaphosphoric acid, used in vitamin C assay, 15–16, 67, 70, 77–8, 80, 87, 90
Methods, overview of, 3
Milk, vitamin D in, 141, 151

Niacin group
  acid-hydrolysable forms of, 155
  bound form of, 13, 155
  references only given for, 4, 155–7
  vitamers listed for, 9
Niacinamide, 9
Nicotinamide, 9, 155
Nicotinamide adenine dinucleotide (NAD), 9, 155
Nicotinamide adenine dinucleotide phosphate (NADP), 9, 155
Nicotinic acid, 9, 155
4-Nitro-1,2-phenylenediamine (NPD), 86, 93
  removal of excess, 93–4
Norit activated carbon, 77, 80, 81
Nutritional surveys, vitamin determination for, 5, 12–13, 18

Official analytical methods, 13
  precision required in, 18–19
Oils and fats, tocopherol determination in, 107–15
Oxidation procedures, thiamine-to-thiochrome, 59–61
Oxygen sensitivity
  vitamin A, 26
  vitamin C, 69–70

Palm oil, tocopherols in, 113
Pantothenic acid, methods/references omitted for, 5
Parsley extract, ascorbic acid in, 72
Peanut oil, retinol protected by, 29
Pragmatic approach, 6

Precision
  concentration effect on, 17
  definition of, 17
Procedure, details necessary for, 4
Pteroylglutamic acid, 12, 158
Purification procedures
  vitamin D compounds, 145–6
  vitamin E compounds, 111
Purpose and scope, of methods
  general discussion, 5–13
  importance of, 3
Pyridoxal (PAL), 10
  determination of, 130, 134–5, 136–7
Pyridoxal 5′-phosphate, 10
Pyridoxamine (PAM), 10
  determination of, 130, 134–5, 136–7
Pyridoxamine 5′-phosphate, 10
Pyridoxine, 10
  determination of, 130, 134–5, 136–7
Pyridoxol (POL), structure of, 10

Quinoxaline derivative, dehydroascorbic acid, 76, 82

Recommended methods
  carotene, 33–49
  vitamin A, 23–31
  vitamin B_1, 51–64
  vitamin C, 66–95
  vitamin E, 97–117
Recovery rates
  carotene, 39, 44, 48
  vitamin A, 31
  vitamin B_1, 63–4
  vitamin B_2, 125
  vitamin B_6, 138
  vitamin C, 73, 83, 95
  vitamin D, 147, 149
Reliability
  carotene methods, 39, 44–5, 48–9
  definition of, 18
  α-tocopherol HPLC method, 105
  vitamin A HPLC method, 31
  vitamin B_1 thiochrome method, 63–4
  vitamin B_2 HPLC method, 125

Reliability—contd.
  vitamin B_6 HPLC method, 138
  vitamin C
    HPLC method, 73
    Sephadex method, 95
  vitamin D HPLC method, 147, 149–50
Repeatability
  carotene methods, 39, 44, 49
  definition of, 17
  α-tocopherol HPLC method, 105
  vitamin A methods, 31
  vitamin B_1 thiochrome method, 64
  vitamin B_2 HPLC method, 125
  vitamin B_6 HPLC method, 138
  vitamin C
    HPLC method, 73
    Sephadex method, 95
  vitamin D HPLC method, 149
Retinaldehyde, 6
  discounted by analytical method, 10
11-*cis*-Retinol, 6
13-*cis*-Retinol, 6
all-*trans*-Retinol, 6
  determination of, 23–31
  α-tocopherol assay affected by, 102
  vitamin A active compounds converted to, 10, 23
Retinyl acetate, 6
Retinyl esters, 6
  determination of, 23–31
Retinyl palmitate, 6
Riboflavin, 8
  HPLC method for, 119–27
Riboflavin adenine dinucleotide, 8
Riboflavin 5'-phosphate, 8
Robustness, analytical method, 18
  checking of, 19
Rye flour, vitamin B_6 compounds in, 137

Sampling procedures, 4, 14–16
  carotene assay, 35–6, 41, 46
  vitamin A assay, 26
  vitamin B_1 thiochrome method, 56–7

Sampling procedures—contd.
  vitamin B_2 HPLC method, 121–2
  vitamin B_6 HPLC method, 132–3
  vitamin C assay, 15–16, 69, 79, 89–90
  vitamin D HPLC method, 144
  vitamin E HPLC methods, 99–100, 109
Saponification techniques
  carotene assay, 36, 42
  vitamin A method, 27
  vitamin D assay, 144
  vitamin E assay, 100
Sephadex DEAE A-25 anion-exchanger, 86, 88, 91
Sephadex method, vitamin C assay, 85–95
Spectrofluorimetric measurements, thiochrome, 62
Spectrophotometric measurements, carotene assay, 38, 43–4, 48
Spherisorb RB ODS columns, 130, 132, 134
Standard deviation, statistical error expressed as, 16, 17–18
Starchy foods
  niacin in, 157
  vitamin C assay for, 93
Statistical data
  general suggestions on, 19
  missing for analytical methods, 4, 19
Supercosil C_8 columns, 25, 29

Tentative methods
  vitamin B_2, 119–27
  vitamin B_6, 129–39
  vitamin D, 141–50
Testolin method, vitamin B_2 assay, 127
5,6,7,8,-Tetrahydrofolic acid, derivatives, 12, 158
Thiamine, analytical method for, 51–64
Thiamine chloride hydrochloride, 7
  vitamin B_1 compounds calculated as, 51, 63

Thiamine diphosphate, 7
Thiamine disulphide, 7
Thiamine esters, 7
  analytical method for, 51–64
Thiamine monophosphate, 7
Thiamine pyrophosphate (TPP), 7
Thiochrome
  fluorimetric measurement of, 62
  thiamine compounds oxidised to, 59–61
α-Tocopherol, recommended method for, 97–105
α-Tocopherol acetate, 11
Tocopherols, 11
  chromatographic separation of, 112, 113
  individual determination of, 107–15
  pre-purification of, 111
  separated by straight phase HPLC, 98
Tocotrienols, 11
  individual determination of, 107–15
  not determinable by recommended method, 97
  separated by straight phase HPLC, 98
Tomato extract, ascorbic acid in, 72
Tryptophan, as niacin precursor, 155

UV absorbance detection, vitamin D compounds, 146, 148–9
UV radiation, vitamin A sensitivity to, 26

Vitamers, 5–13
  interconversion of, 5, 6
  methods not selective for, 5, 6, 9–10, 12
Vitamin A group
  IU equivalents, 23, 24–5
  recommended method for, 23–31
  vitamers listed for, 6

Vitamin activity, defined, 5
Vitamin B_1 group
  recommended method for, 51–64
  vitamers listed for, 7
Vitamin B_2 group
  recommended method for, 119–27
  tentative method for, 119–27
  vitamers listed for, 8
Vitamin B_6 group
  separation of vitamers of, 130
  tentative method for, 129–39
  vitamers listed for, 10, 129
Vitamin B_{12}, methods/references omitted for, 5
Vitamin C group
  recommended methods
    HPLC method, 66–74
    modified Deutsch and Weeks fluorimetric method, 76–84
    Sephadex method, 85–95
  vitamers listed for, 10
Vitamin D group
  method non-selective of vitamers of, 12
  tentative method for, 141–50
  vitamers listed for, 11
Vitamin D_2, see Ergocalciferol
Vitamin D_3, see Cholecalciferol
Vitamin E group
  recommended method for α-tocopherol only, 12, 97–106
  recommended methods, 97–115
  vitamers listed for, 11
Vitamin K, methods/references omitted for, 5

Wheat flour extract, riboflavin in, 126

Xanthopyhlls, carotenes separated from, 38, 40, 42, 47